Optical WDM Networks

IEEE Press
445 Hoes Lane
Piscataway, NJ 08854

Optical WDM Networks

Concepts and Design Principles

Jun Zheng
Hussein T. Mouftah
School of Information Technology and Engineering
University of Ottawa

IEEE PRESS

A JOHN WILEY & SONS, INC., PUBLICATION

Published simultaneously in Canada.

For general information on our other products and services please contact our Customer Care Department within the U.S. at 877-762-2974, outside the U.S. at 317-572-3993 or fax 317-572-4002.

Wiley also publishes its books in a variety of electronic formats. Some content that appears in print, however, may not be available in electronic format.

Library of Congress Cataloging-in-Publication Data is available.

ISBN 0-471-67170-3

10 9 8 7 6 5 4 3 2 1

To our parents and family

Contents

Preface

In the past decade, we have witnessed the huge success and spectacular growth of the Internet, which has led to the explosion of Internet traffic and has therefore imposed a huge bandwidth demand on its underlying telecommunications infrastructure. To meet the unprecedented demand for bandwidth, fiber optics has brought about a bandwidth revolution in telecommunications networks. At the core of this revolution, optical fiber has proved to be an excellent physical transmission medium because of its huge transmission bandwidth (nearly 50 terabits) as well as a number of other advantages over traditional transmission media, such as low signal attenuation (about 0.2 dB/km), low error bit rate (typically 10^{-12}), low signal distortion, low power requirement, low space requirement, and low cost. In this revolution, the emergence of wavelength division multiplexing (WDM) technology is a new milestone. WDM allows multiple optical signals to be transmitted independently and simultaneously in multiple optical channels or wavelengths over a single fiber, each operating at a very high rate of a few gigabits per second (Gbps), and can thus more efficiently exploit the usable bandwidth inherent in optical fibers. With recent advances in enabling technologies, WDM systems capable of supporting up to 160 channels at 10 Gbps are commercially available and products with more optical channels are expected to come into the market soon. Therefore, WDM has been widely considered a technology of choice for meeting the huge bandwidth demand in telecommunications networks. Optical networks using WDM technology have become the most promising network infrastructure for next-generation telecommunications networks, not only for wide-area networks but also for metropolitan area networks and local area networks.

Although WDM technology is currently being deployed by many network providers mostly for point-to-point transmission, a large effort from academia, industry, and standardization organizations has been and is being made to enable the transition of WDM from a point-to-point transmission technology toward a networking technology. Because of this effort, significant research and development progress has been made over the last few years. Advanced optical devices, such as erbium-doped fiber amplifiers (EDFAs), optical cross-connects (OXCs), and optical add/drop multiplexers (OADMs) have become commercially available. A number of experimental prototypes and testbeds have already been and are currently being developed and built by many network and service providers. Although there are still a number of challenges for the vision of all-optical networks to be commercially realized, optical WDM networks are indeed coming into the marketplace rapidly.

The purpose of this book is to provide an introduction to basic concepts, major issues, and effective solutions for wavelength-routed WDM networks. Distinguished from other books in optical networks, this book focuses primarily on the networking aspects of such networks and highlights the fundamental concepts and design principles. In addition, the state-of-the-art developments and technologies are introduced. The book is organized into eight chapters and covers the most important networking aspects, such as network control architecture, routing and wavelength assignment, virtual topology configuration and reconfiguration, network control and management, optical-layer protection and restoration, and IP over WDM. To help readers to better understand the contents of each chapter, a number of examples are given to illustrate the concepts, problems, and solutions and a number of problems are included at the end of each chapter. In addition, a list of extensive references is provided in each chapter for those readers who seek a deeper exploration.

This book is intended for graduate students and academic researchers as an introduction to the design of wavelength-routed WDM networks and further their research work. In particular, it is intended as a textbook for graduate students in electrical engineering, computer engineering, or computer science. It can be used as a textbook for a graduate course on optical networks or as a supplementary textbook for a graduate or senior undergraduate course on telecommunications networks, data communication networks, or computer networks. This book may also be of interest to network engineers, designers, planners, operators, and managers in the area of electrical and computer engineering, who would like to learn more about optical networks.

Jun Zheng
Hussein T. Mouftah
University of Ottawa

Acknowledgments

This book would not have been possible without support, encouragement, and contribution from many people in their different ways.

We are grateful to the editors and production group at Wiley-IEEE Press for their enthusiastic support for this project. The publication of this book would not have been possible without the vision and foresight of our Senior Acquisitions Editor, Catherine Faduska, and Wiley-IEEE Press. Many thanks go to Catherine Faduska for her efforts throughout the entire process of this project, to our Project Editor, Anthony VenGraitis, for coordinating the production of the book, and to our Managing Editor, Danielle Lacourciere, and Copy Editor, Suan Adams, for carefully copyediting the whole manuscript.

We would also like to thank the anonymous reviewers for their time in carefully reviewing our book as well as their constructive and valuable comments.

Special thanks are given to a number of friends and colleagues for their support and encouragement throughout the writing of this book. We will not forget their gentle reminding message that there are more beautiful things in life beyond working and writing.

Last but definitely most, Jun Zheng would like to express his deep gratitude to his parents for their invaluable love, understanding, and support during the writing of the book. Hussein T. Mouftah is deeply grateful to his students. He would like to thank his wife, Ebtisam, and his daughters, Maye, Nadine, and Nermeen, for their support, understanding, and sacrifice of time spent while working on this project.

Chapter 1

Introduction

1.1 Optical Networks: A Brief Picture

The rapid evolution of telecommunications networks is always driven by ever-increasing user demands for new applications as well as continuous advances in enabling technologies. In the past ten years, we have witnessed the huge success and explosive growth of the Internet, which has attracted a large number of users surging into the Internet. Individual users are using the Internet insatiately for information, communication, and entertainment, while enterprise users are increasingly relying on the Internet for their daily business operations. As a result, Internet traffic has experienced an exponential growth in the past ten years, which is consuming more and more network bandwidth. On the other hand, the emergence of time-critical multimedia applications, such as Internet telephony, video conferencing, video on demand, and interactive gaming, is also swallowing up a large amount of network bandwidth. All these facts are imposing a tremendous demand for bandwidth capacity on the underlying telecommunications infrastructure.

To meet the unprecedented demand for bandwidth capacity, a bandwidth revolution has taken place in telecommunications networks with the introduction of fiber optics. As the core of this revolution, optical fibers have proved to be an excellent physical transmission medium for providing huge bandwidth capacity. Theoretically, a single single-mode fiber has a potential bandwidth of nearly 50 terabits per second (Tbps), which is about four orders of magnitude higher than the currently achievable electronic processing speed of a few gigabits per second (Gbps) [1]. Apart from the huge bandwidth capacity, optical fibers also have a number of other significant characteristics, such as low signal attenuation (about 0.2 dB/km), low error bit rate (typically, 10^{-12}), low signal distortion, low power requirement, low space requirement, and low cost [1–3]. However, because of the limit of the electronic processing speed, it is unlikely that all the bandwidth of an optical fiber will be exploited by using a single high-capacity optical channel or wavelength. For this reason, it is desirable to find an effective technology that can efficiently exploit the huge potential bandwidth capacity of optical fibers. The emergence of wavelength division multiplexing (WDM) technology has provided a practical solution to meeting this challenge. With WDM technology, multiple optical signals can be transmitted simultaneously and independently in different optical channels over a single fiber, each at a rate of a few gigabits per second, which significantly increases the usable bandwidth of an optical fiber. Recently, commercial WDM systems with up to 160 OC-192 (10 Gbps) channels have been announced [4]. In addition to the increased usable bandwidth of an optical fiber, WDM also has a number of other advantages, such as reduced electronic processing cost, data transparency, and efficient failure handling [1]. As a result, WDM has become a technology of choice for meeting the tremendous bandwidth demand in the telecommunications infrastructure. Optical networks employing WDM technology have been widely considered a promising network infrastructure for next-generation telecommunications networks and optical Internet [4–10].

Today, WDM technology is being deployed in various types of telecommunications networks. However, the deployment is mainly for point-to-point transmission [1]. All the routing and switching functions are still performed electronically at each network node. Optical signals must go through opto-electronic (O/E) and electro-optical (E/O) conversion at each intermediate node as they propagate along an end-to-end path from one node to another node. Consequently, a network node may not be capable of processing all the traffic carried by all its input signals, including the traffic intended for the node as well as the traffic that is just passing through the node to other network nodes, causing an electronic bottleneck. For example, a node with four input fiber links and 16 wavelengths at 2.5 Gbps on each fiber link must handle a maximum data rate of 160 Gbps, which is normally beyond the electronic processing capability of the node. To overcome the electronic bottleneck, it is desirable to incorporate optical routing and switching functions at each network node in order to bypass those signals that carry traffic not intended for the node directly in the optical domain. With the advent of reconfigurable optical devices, such as optical add-drop multiplexers (OADMs) and optical cross-connects (OXCs) [1–2], this has become possible. At the time of this writing, OADM and OXC products are commercially available from several product vendors and more advanced products are expected to become available in the near future [4]. A number of WDM networks have been tested and are being tested in the United

States, Europe, and many other countries. It is expected that WDM technology will be widely deployed in all types of telecommunications networks, not only in wide area networks or backbone networks, but also in metropolitan and local area networks [11–15].

1.2 WDM Technology

Wavelength division multiplexing (WDM) is an optical multiplexing technology for exploiting the huge bandwidth capacity inherent in optical fibers. Conceptually, it is similar to frequency division multiplexing (FDM) that has already been used in radio communication systems for over a century. The basic principle is to divide the huge bandwidth of an optical fiber into a number of nonoverlapping subbands or optical channels and transmit multiple optical signals simultaneously and independently in different optical channels over a single fiber, each signal being carried by a single wavelength.

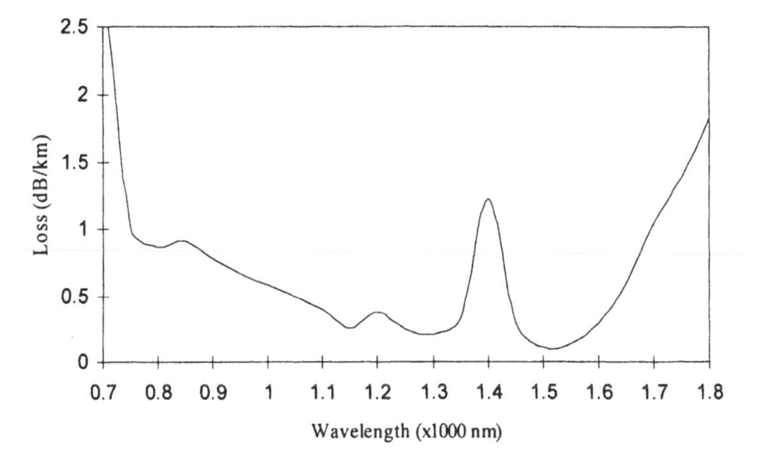

Figure 1.1 Transmission spectrum of optical fibers.

Figure 1.1 shows the transmission spectrum of an optical fiber. In this spectrum, there are two low-attenuation areas. One is centered at 1300 nanometers (nm) and the other at 1500 nm. Both areas have a range of about 200 nm with an attenuation loss less than 0.5 dB/km. Theoretically, these two areas can provide a total amount of 50 Tbps low-attenuation transmission bandwidth. However, because the maximum rate at which an end device can access an optical channel is limited by its electronic processing speed, it is technically impossible to take advantage of all the bandwidth of an optical fiber by using a single optical channel or wavelength. With WDM, the huge bandwidth is carved up into a number of optical channels with each channel operating at any feasible rate, say, a few gigabits per second to be compatible with current electronic processing speeds. Accordingly, a single fiber is theoretically capable of supporting over 1000 optical channels or wavelengths at a few gigabits per second.

Figure 1.2 shows a block diagram of a basic WDM transmission system. The network medium can be a simple fiber link, a passive star coupler, or any type of optical network. The transmitter consists of a laser and a modulator. The laser is the light source, which generates an optical carrier signal at either a fixed wavelength or a tunable wavelength. In the modulator, the carrier signal is modulated by an electronic signal and is then sent to the multiplexer (MUX). The multiplexer combines multiple optical signals on different wavelengths at its input ports into a single optical signal, which is transmitted to a common output port or optical fiber. The demultiplexer (DMUX) uses optical filters to separate the optical signal received on the input port into multiple optical signals on different wavelengths, which are then sent into the receivers. The receiver consists of a detector (e.g., photodiode) that can convert an optical signal to an electronic signal. The optical amplifiers are used to maintain the power strength of an optical signal at appropriate locations in the transmission system.

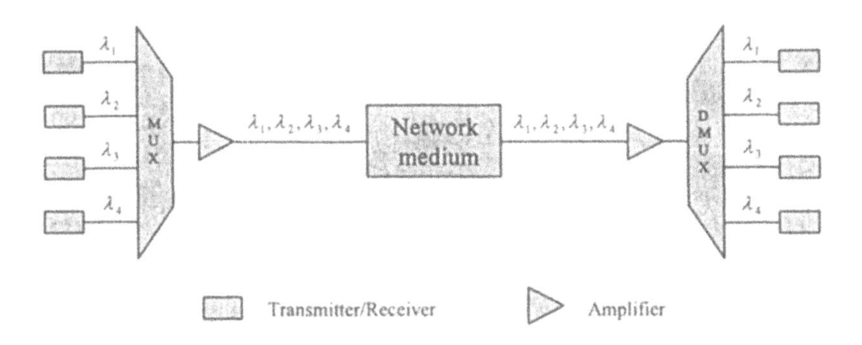

Figure 1.2 A WDM transmission system.

Figure 1.3 illustrates the WDM of a transmission bandwidth of 250 Gbps. The transmission bandwidth is divided into a number of optical channels, each having a smaller bandwidth at a particular wavelength. Note that it is important to have sufficient spacing between the wavelengths of adjacent channels to avoid interchannel cross talk or interference caused by spectrum overlapping and transmission imperfections. Accordingly, the size of the frequency spacing should be no less than the bandwidth of each channel. The density of channels depends on both the channel bandwidth and the frequency spacing. For example, if the bandwidth of each channel is 2.5 GHz and the frequency spacing is also 2.5 GHz, there are a total of 100 channels available over a single fiber, resulting in a total transmission capacity of 250 Gbps, as shown in Figure 1.3(a). This is an extreme case in which there is no spacing between the wavelengths of two adjacent channels. In this case, it is practically very difficult to prevent adjacent signals from interfering with each other because of the impairments occurring in transmission, such as dispersion and nonlinearity. If the channel bandwidth becomes 10 GHz and the frequency spacing becomes 20 GHz, only 13 channels are available, resulting in a transmission capacity of 130 Gbps, as shown in Figure 1.3(b). Therefore, given the bandwidth of each channel, the frequency spacing determines the density of optical channels.

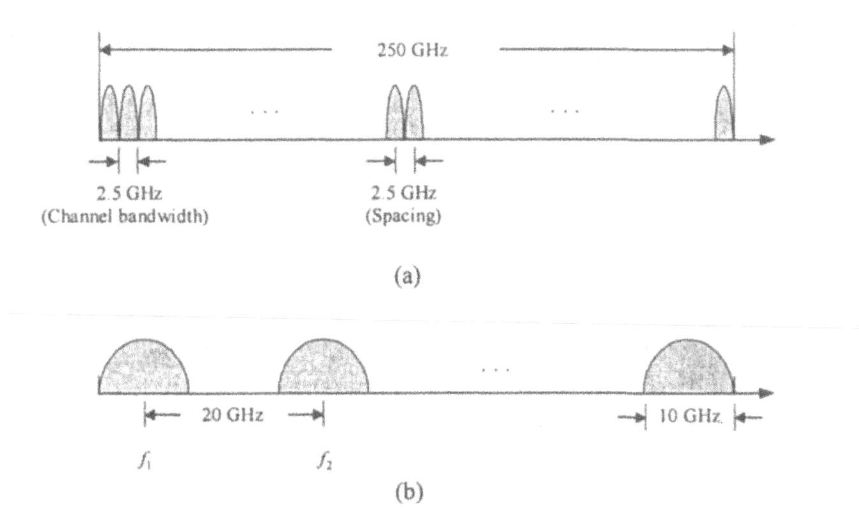

Figure 1.3 Illustration of WDM.

On the other hand, an optical channel provides a bit rate of a few gigabits per second. To make efficient use of this bandwidth, it is often necessary to share a single optical channel among multiple low-bit-rate channels. The most common way to implement this is to use time division multiplexing (TDM) technology [1]. Although a single fiber can theoretically support a number of optical channels or wavelengths, the development of more wavelengths over a single fiber depends on advances in enabling technologies as well as the commercial availability of WDM devices.

Fortunately, the last few years have seen a rapid growth of WDM transmission systems. Five years ago, commercially available WDM systems could only offer up to 32 optical channels at 2.5 Gbps over a single fiber. At the time of this writing, however, 160 channels at 10 Gbps (OC-192) have been announced as commercially available. It is expected that WDM systems supporting more channels or wavelengths will come into the marketplace in the next few years.

1.3 WDM Network Architectures

WDM networks can be classified into two broad categories: broadcast-and-select WDM networks and wavelength-routed WDM networks.

1.3.1 Broadcast-and-Select WDM Networks

A WDM network that shares a common transmission medium and employs a simple broadcasting mechanism for transmitting and receiving optical signals between network nodes is referred to as a broadcast-and-select WDM network. The most popular topologies for a broadcast-and-select WDM network are the star topology and the bus topology, as shown in Figure 1.4 and Figure 1.5, respectively. In the star topology, a number of nodes are connected to a passive star coupler by WDM fiber links. Each node has one or more optical transmitters and receivers, which can be either fixed-tuned or tunable. A node transmits its signal on an available wavelength. Different nodes can transmit their signals on different wavelengths simultaneously and independently. The star coupler receives and combines all the signals and broadcasts them to all the nodes in the network. To receive a signal, a node tunes one of its receivers to the wavelength on which the signal is transmitted.

In the bus topology, a number of nodes are connected to a bus through 2×2 couplers (see Section 2.3) by WDM fiber links. Each node transmits its signal to the bus on an available wavelength through a coupler and receives a signal from the bus through another coupler. In the transmitting coupler only one of the output ports is used, and in the receiving coupler only one of the input ports is used. Both the star and bus topologies use optical couplers. A star coupler can be made out of 2×2 couplers. The two topologies differ in the number of couplers used and in the manner in which the couplers are connected. In most cases, the star topology has proven to be a better choice for many types of networks [1].

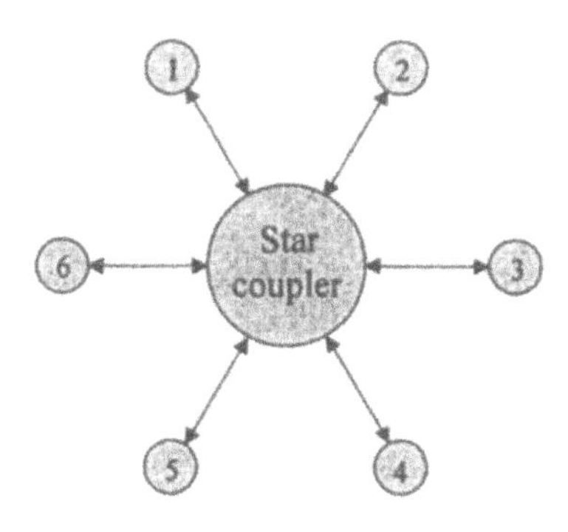

Figure 1.4 A star topology.

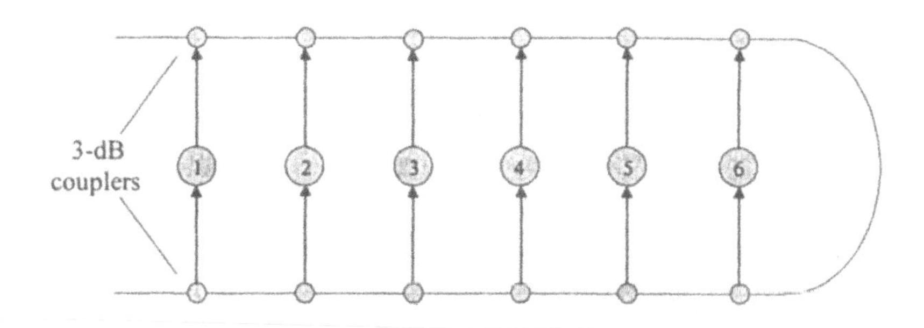

Figure 1.5 A bus topology.

The advantage of a broadcast-and-select WDM network lies in its simplicity and broadcasting capability. However, this type of network needs a large number of wavelengths because the wavelengths cannot be reused in the network. A wavelength can only be used by one node at a given time. As a result, the network is not scalable to the number of nodes in the network. On the other hand, because the transmitted power from a node is split among all the nodes in the network, each node can only receive a very small fraction of the transmitted power. Accordingly, this type of network cannot span a long distance and is most suitable for deployment in local area networks (LANs) or metropolitan area networks (MANs). Because broadcast-and-select WDM networks are not the focus of this book, the readers are referred to [1–2] for more discussion.

1.3.2 Wavelength-Routed WDM Networks

A WDM network that employs wavelength routing to transfer data traffic is referred to as a wavelength-routed WDM network. A wavelength-routed WDM network typically consists of routing nodes interconnected by point-to-point WDM fiber links in an arbitrary mesh topology. Each routing node employs a set of transmitters and receivers for transmitting signals to and receiving signals from fiber links and an optical cross-connect (OXC) or wavelength cross-connect (WXC) to route and switch different wavelengths from an input port to an output port. Each fiber link operates in WDM and supports a certain number of optical channels or wavelengths. An access node can be connected to a routing node, which is used as an interface between the optical network and the electronic client networks. At the source side, an access node performs traffic aggregation and E/O conversion functions. At the destination side, traffic deaggregation and O/E conversion are performed. In the context of this book, an access node and its associated routing node are collectively referred to as a network node or simply a node unless otherwise stated. The architecture of a wavelength-routed WDM network is shown in Figure 1.6.

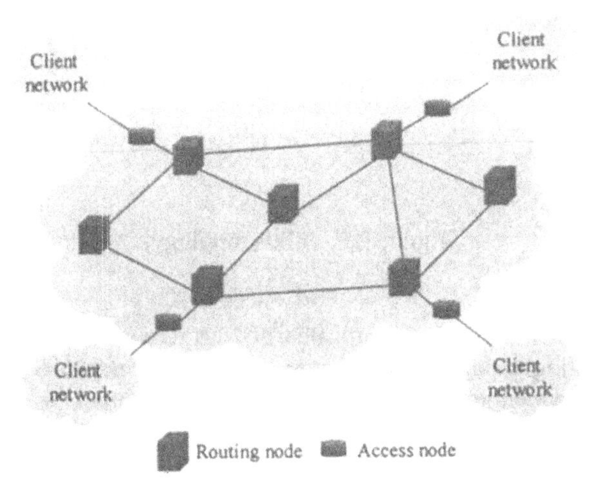

Figure 1.6 A wavelength-routed WDM network.

A wavelength-routed WDM network is a circuit-switched network in which a pair of network nodes communicates through an end-to-end optical connection that may consist of one or more all-optical connections called lightpaths. A lightpath is a unidirectional all-optical connection between a pair of network nodes, which may span multiple fiber links and use one or multiple wavelengths without undergoing any O/E and E/O conversion at each intermediate node. Two lightpaths cannot share the same wavelength on a common fiber link, which is referred to as the wavelength-distinct constraint. However, two lightpaths can use the same wavelength on different fiber links, which is referred to as the wavelength-reuse property. In the absence of any wavelength conversion capability, a lightpath must use the same wavelength on all the links it spans, which is known as the wavelength-continuity constraint. This constraint is unique to wavelength-routed WDM networks and makes such networks different from conventional circuit-switched networks. Because of this constraint, network performance in terms of wavelength utilization and blocking probability would be largely degraded. For this reason, it is desirable to eliminate the wavelength-continuity constraint in order to improve network performance. This can be achieved by deploying wavelength converters at network nodes to provide wavelength conversion capability in the network.

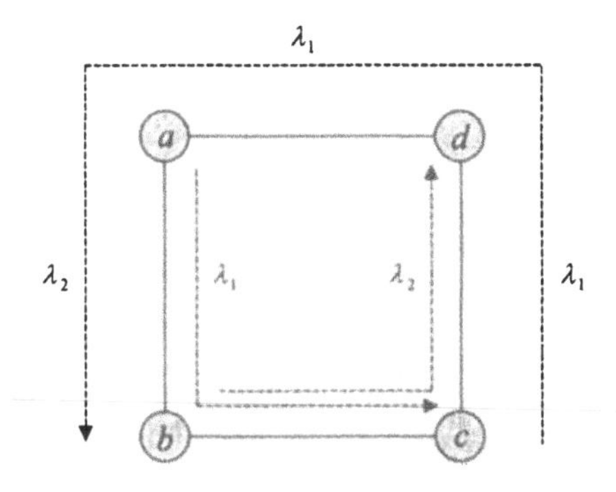

Figure 1.7 Effect of wavelength conversion.

To see how wavelength conversion helps to improve wavelength utilization and blocking probability, let us examine the example shown in Figure 1.7. Suppose that the network has four nodes with two wavelengths on each fiber link. There are three connection requests for a connection from node *a* to node *c*, node *b* to node *d*, and node *c* to node *b*, respectively. It is not difficult to see that all three connections cannot be established simultaneously without using any wavelength converter in the network. For example, if the connection from node *a* to node *c* is established with wavelength *1* (λ_1) on path (*a*-*b*-*c*) and the connection from node *b* to node *d* is established with wavelength *2* (λ_2) on path (*b*-*c*-*d*), no connection can be established from node *c* to node *d* because there is no wavelength available on path (*c-b*) and there is no common wavelength available on path (*c-d-a-b*). However, if node *a* has a wavelength converter that can convert wavelength *1* to wavelength *2*, a connection can thus be established from node *c* to node *d* with wavelength *1* on path (*c-d-a*) and wavelength *2* on path (*a-b*).

A wavelength-routed WDM network can be classified into the following different categories:

- Nonreconfigurable and reconfigurable
- Single hop and multiple hop
- Wavelength selective and wavelength convertible
- Single fiber and multiple fiber

Nonreconfigurable networks

A nonreconfigurable network is also referred to as a static network, in which no switches are employed in the routing nodes. The network configuration does not change once the network is configured. The set of lightpaths established in the network are fixed and are determined at the time of network design. This type of network can consist entirely of passive optical devices, such as passive couplers, multiplexers, and demultiplexers, and thus is more reliable than reconfigurable networks.

Reconfigurable networks

A reconfigurable network employs switches in the routing nodes, which can route a wavelength on an input port to a wavelength on a different output port. Accordingly, the set of lightpaths established in the network can be dynamically reconfigured by changing the statuses of the switches in the routing nodes. This book focuses on reconfigurable networks.

Single-hop networks

A single-hop network is also referred to as an all-optical network, in which data traffic is transferred from one node to another node through a single lightpath without undergoing any intermediate O/E and E/O conversion. Accordingly, a single-hop network can provide the best quality of service to its end users. Because of the limit in network resources (e.g., wavelengths), however, a single-hop network may not be able to achieve the maximum network throughput.

Multiple-hop networks

In a multiple-hop network, an end-to-end connection consists of a tandem of lightpaths. At the end nodes of each lightpath, data traffic must go through O/E and E/O conversion, which would introduce additional conversion delay. Because of the O/E and E/O conversion, congestion may also occur in a routing node, which may further introduce queuing delay. For this reason, a multiple-hop network may not be able to provide the best quality of service (e.g., end-to-end delay) to its end users. However, it provides a higher level of flexibility and can thus provide better network performance in terms of resource utilization, blocking probability, and network throughput.

Wavelength-selective networks

In a wavelength-selective network, no wavelength conversion capability is provided at any routing node. A lightpath must use the same wavelength on all the links it traverses, which is referred to as the wavelength-continuity constraint. Because of this constraint, the network performance in terms of wavelength utilization, blocking probability and network throughput would be largely affected. In this book, we assume no wavelength conversion in all routing nodes and focus on wavelength-selective networks unless otherwise stated.

Wavelength-convertible networks

In a wavelength-convertible network, wavelength converters are deployed in the routing nodes to overcome the wavelength-continuity constraint. A wavelength converter is an optical device that can convert one wavelength to another wavelength in the optical domain. We will introduce wavelength converters in more detail in Section 2.10. With wavelength converters, the wavelength-continuity constraint can be alleviated or even eliminated. As a result, a wavelength-convertible network can achieve better wavelength utilization and smaller blocking probability than a wavelength-selective network.

Single-fiber networks

In a single fiber network, there is only one fiber pair between each pair of network nodes. Network performance is limited by the number of available wavelengths on each fiber link.

Multiple-fiber networks

A multiple fiber network uses multiple fiber pairs between each pair of networks nodes to provide more wavelengths and larger transmission bandwidth. As a result, network performance can be significantly improved. A multiple-fiber network with no wavelength conversion is equivalent to a single-fiber network with limited wavelength conversion [1][16].

From the viewpoint of a layered network hierarchy, the deployment of WDM technology introduces a new layer in the layered hierarchy, called the optical layer, as illustrated in Figure 1.8. The optical layer consists of a set of lightpaths established on the physical topology of the network. It involves a variety of networking functions, such as wavelength routing and switching, wavelength multiplexing and demultiplexing, and wavelength adding and dropping, and provides circuit-switched lightpath service to its higher layers, such as SONET/SDH, ATM, and IP. According to International Telecommunications Union-Telecommunication Standard Sector (ITU-T) Recommendation G.872, the optical layer itself consists of the following three sublayers:

- An optical channel layer
- An optical multiplex section layer
- An optical transmission section layer

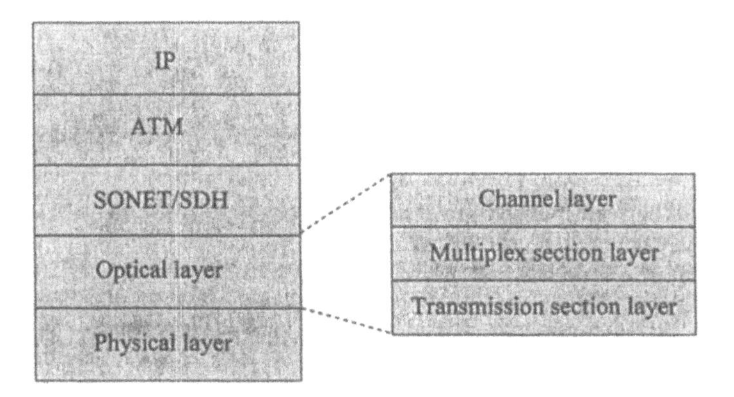

Figure 1.8 Layered network hierarchy.

The optical channel layer provides wavelength routing and switching functions to establish end-to-end optical connections between network nodes for transparent data delivery. The optical multiplex section layer is responsible for the aggregation of multiple signals. The optical transmission section layer handles the transmission of optical signals on different types of optical fibers such as single-mode and multimode fibers. The optical layer and the layered hierarchy will be discussed in more detail in Chapter 7.

In addition to the high bandwidth capacity provided and the circuit-switched characteristic, the optical layer has several other significant characteristics, including transparency, wavelength reuse, and reliability.

Transparency

Transparency refers to the capability that a lightpath can carry data traffic at a variety of bit rates and in a variety of protocol formats. Because a lightpath is an all-optical end-to-end connection from one node to another node without any O/E and E/O conversion, data traffic can be transferred over a lightpath entirely in the optical domain at any bit rate and in any protocol format. This makes the optical layer capable of supporting a variety of higher layers simultaneously. They can carry not only SONET/SDH traffic but also ATM cells or IP packets, all at different bit rates.

Wavelength reuse

Wavelength reuse refers to the capability that two lightpaths can use the same wavelength on different links. With this capability, the network can make efficient use of wavelength resources and can thus significantly improve network performance. Wavelength reuse also makes wavelength-routed WDM networks more scalable than broadcast-and-select networks.

Reliability

Reliability refers to the capability that the network recovers from network failures. In the event of a failure, a backup path can be automatically provisioned for a disrupted lightpath. This provides a high degree of reliability at the optical layer of the network. On the other hand, many optical devices (e.g., passive multiplexers and demultiplexers) are passive devices, which are inherently reliable and thus provide another degree of reliability at the optical layer.

The significant characteristics of the optical layer have made wavelength-routed WDM networks a promising network infrastructure for next-generation telecommunications networks.

1.4 Focus of This Book

This book is intended as an introduction to wavelength-routed WDM networks. The focus is primarily on the fundamental concepts and principles related to the networking aspects of the design of wavelength-routed WDM networks. The major topics covered in this book include routing and wavelength assignment (RWA), virtual topology design and reconfiguration, lightpath control and management, network protection and restoration, and IP over WDM for next-generation optical Internet, which are briefly described as follows.

1.4.1 Routing and Wavelength Assignment

Routing and wavelength assignment (RWA) is one of the key issues in wavelength-routed WDM networks [1]. As discussed in Section 1.3.2, such networks use all-optical connections called lightpaths to transfer data traffic between network nodes and a lightpath must first be established between a pair of nodes before data traffic can be transferred. To establish a lightpath, the network must first decide on a physical route and then assign an available wavelength on each link of the decided route, which is referred to as the RWA problem. An algorithm that is used to perform RWA is called the RWA algorithm. Because of the limit in the number of wavelengths available on each fiber link as well as the wavelength constraints imposed on the network, RWA becomes very important in the design of wavelength-routed WDM networks and an RWA algorithm plays a crucial role in achieving good network performance. For this reason, the RWA problem has been widely studied for many years and a lot of research results can be found in the literature [17].

In general, the RWA problem has a different objective for different types of network traffic. For static traffic, the objective is to accommodate a given set of connection requests using a minimum number of wavelengths or to establish as many lightpaths as possible for a given number of wavelengths. For dynamic traffic, the objective is to route and assign wavelengths to establish lightpaths in a manner that can minimize the blocking probability in the network or maximize the network throughput in terms of the number of lightpaths established in the network. These have presented a big challenge for network designers. This book will discuss both static and dynamic RWA problems.

1.4.2 Virtual Topology Design and Reconfiguration

Virtual topology design is another important issue in the design of wavelength-routed WDM networks. A virtual topology consists of a set of lightpaths established over the physical topology of the network and forms an optical layer between the physical layer and the higher layers. This optical layer is protocol transparent and can thus support different network services at its higher layers, such as SONET/SDH, ATM, and IP. Because of the limit in the number of wavelengths available on a fiber link as well as other network resources, such as transmitters and receivers, it may not be possible to establish a lightpath between each pair of nodes in the network. For this reason, virtual topology design becomes very important. Given the physical topology, the network constraints, and the traffic demand between each pair of network nodes, the problem of establishing a set of lightpaths over the physical topology to accommodate the traffic demand and at the same time optimizing the network performance is referred to as the virtual topology design problem. This problem can be typically formulated as a mixed-integer linear program (MILP), which is known to be NP-hard and is therefore computationally intractable. A practical and effective approach to this problem is to use heuristics to obtain approximate solutions. For this reason, the design of an efficient heuristic algorithm becomes a great concern.

On the other hand, the traffic demand and physical topology in the network may change because of various practical factors. In response to the traffic demand changes and physical topology changes in the network, a virtual topology may need to be reconfigured in order to maintain optimal network performance. Typically, given the physical topology, the original virtual topology, and the new traffic demand, the reconfiguration problem is to design a new virtual topology that can optimize network performance and minimize the service disruption incurred during the reconfiguration. In general, the reconfiguration problem is closely related to the virtual topology design problem. Both virtual topology design and reconfiguration have been widely studied in the literature. A good survey of virtual topology design algorithms can be found in [18]. This book will discuss both virtual topology design and reconfiguration.

1.4.3 Lightpath Control and Management

For dynamic traffic, the network must have an efficient lightpath control mechanism to dynamically establish a lightpath for each connection request. In general, lightpath control can be either centralized or distributed. Under centralized control, there is a centralized controller in the network, which maintains global network state information and is responsible for lightpath establishment on behalf of all network nodes. Obviously, centralized control is relatively simple to implement and works well for static traffic. However, it is not scalable and reliable and may become the bottleneck of the network. A failure with the centralized controller may result in the breakdown of the entire network. For these reasons, centralized control is considered unsuitable for large networks with dynamic traffic. Under distributed control, all connection requests are processed at different network nodes simultaneously and each node makes its decisions independently based on the network state information it maintains, which can be either local or global. Compared with centralized control, distributed control improves the scalability and reliability of the network and is therefore highly preferred for large networks with dynamic traffic. However, distributed control increases the difficulty and complexity in lightpath control, which also presents a big challenge in network design.

Distributed lightpath control involves not only the routing and wavelength assignment problems but also the wavelength reservation problem. Any of these problems may have a significant impact on the network performance. For this reason, distributed lightpath control has received a lot of attention in recent years [19–20]. A variety of routing algorithms, wavelength assignment algorithms, and wavelength reservation protocols has been proposed in the literature for distributed lightpath control, which are the focus of Chapter 5.

1.4.4 Optical Layer Survivability

Network survivability has been a great concern for all types of high-speed telecommunications networks. In wavelength-routed WDM networks, a fiber provides a number of optical channels or wavelengths to carry data traffic, each operating at a very high rate of several gigabits per second. A single network failure such as a fiber cut may cause the disruption of all the lightpaths that traverse the failed fiber link and can therefore lead to a large amount of data loss in the network. This would largely degrade and even disrupt the network services. For this reason, survivability is of particular importance in such networks. To guarantee network services, the network must provide effective survivability capabilities to survive different types of network failures (e.g., a fiber cut or a node fault). From the perspective of layered architecture, network survivability can be provided at different layers of the network, such as IP, ATM, SONET, and the optical layer. Although the higher IP, ATM, and SONET layers may have their own protection and restoration mechanisms, it is still attractive to provide survivability at the optical layer because of a number of advantages, such as fast restoration, efficient resource utilization, and protocol transparency. For this reason, optical layer survivability has been a hot topic for many years and is continuing to receive a lot of attention [21–25].

In general, there are two basic survivability paradigms for the optical layer: predetermined protection and dynamic restoration [5-6]. In predetermined protection, spare network resources are reserved at the time of network design or connection establishment for protection against network failures. In the event of a network failure, the disrupted network services are recovered by using the reserved network resources. Obviously, predetermined protection is fast in service recovery as the reserved network resources are dedicated to network failures. However, it is inefficient in resource utilization because the reserved network resources cannot be used for other connections. In dynamic restoration, no spare resources are reserved in advance for restoration purposes. The network must dynamically discover spare network resources available in the network to recover the disrupted network services after network failures occur. Obviously, dynamic restoration is efficient in resource utilization but slow in service recovery. For these reasons, efficient resource utilization and fast service recovery become a great concern with predetermined protection and dynamic restoration, respectively. We will discuss both predetermined protection and dynamic restoration in Chapter 6.

1.4.5 IP over WDM

The emergence of the Internet and its supported applications based on the Internet Protocol (IP) has opened up a new era in telecommunications. In the past two decades, we have witnessed the huge success and explosive growth of the Internet, which have been driving the demand for larger bandwidth in the underlying transport infrastructure of the Internet. With new time-critical multimedia applications such as Internet telephony, video conferencing, and video on demand becoming feasible and pervasive in the Internet, this bandwidth demand will continue to grow at a rapid rate. It has been widely believed that IP is going to be the common traffic convergence layer in telecommunications networks and that IP traffic will become the dominant traffic in the future. In parallel with the growth of the Internet, optical technology has also seen its rapid development in the past ten years. The emergence of WDM technology has provided an unprecedented opportunity to dramatically increase the bandwidth capacity of telecommunications networks. Currently, there is no other technology on the horizon that can meet the huge demand for bandwidth in the Internet transport infrastructure more effectively than WDM technology [8]. For this reason, IP over WDM has been envisioned as the most promising network architecture for the next-generation optical Internet. To support IP over WDM, a number of technical issues must be addressed, such as IP over WDM layered models, service models, interconnection models, control planes for IP over WDM, and network survivability. These issues have received a lot of attention in recent years, not only in academia but also in industry and standards organizations. A lot of research efforts including standardization activities have been carried out and are still going on to realize the vision of IP over WDM in the next-generation optical Internet [26–28].

1.5 Outline of This Book

This chapter serves as an introduction to the whole book. The background and historical perspectives of WDM networks are briefly given. WDM technology and WDM network architectures are introduced. The fundamental concepts related to wavelength-routed WDM networks are also introduced.

Chapter 2 gives an introduction of the fundamentals of major optical devices that are used in WDM networks. These WDM network devices include optical fibers, couplers, amplifiers, transmitters, receivers, multiplexers, demultiplexers, optical add/drop multiplexers, optical switch/cross-connects, and wavelength converters. To help those readers who have no background in fiber optics, we try to minimize the optics involved in these devices and focus on the basic functions and characteristics.

Chapter 3 is devoted to the routing and wavelength assignment (RWA) problem. The concept of the RWA problem is introduced, and the objectives of the RWA problem for both static traffic and dynamic traffic are described. Both static RWA and dynamic RWA are discussed, and effective solutions are presented, including integer linear programming (ILP) formulations for static RWA and a variety of RWA algorithms for dynamic RWA. The RWA fairness problem is discussed, and effective RWA fairness methods are introduced. In addition, this chapter also discusses the wavelength rerouting problem. The basic lightpath migration operations and rerouting schemes are introduced, and a well-known wavelength rerouting algorithm is presented. The focus is primarily on wavelength-selective networks with the wavelength-continuity constraint.

Chapter 4 focuses on the virtual topology design and reconfiguration problem. The virtual topology design problem and subproblems are first described, and the basic concepts and network constraints related to the virtual topology design are then introduced. The virtual topology design problem is formulated as a mixed-integer linear program (MILP), and an exact formulation of the problem is presented. This problem is known to be NP-hard and is therefore computationally intractable. A variety of heuristic algorithms that have already been proposed in the literature are presented, including those for regular topology design, predetermined topology design, and arbitrary topology design. Moreover, the virtual topology reconfiguration problem is also discussed, including reconfiguration for traffic changes and reconfiguration for topology changes.

Chapter 5 is dedicated to distributed lightpath establishment. The major issues and concerns with distributed lightpath establishment are discussed, and basic routing and wavelength reservation paradigms are introduced. Moreover, a variety of routing algorithms, wavelength assignment algorithms, and wavelength reservation protocols that have already been proposed for distributed lightpath establishment are presented.

Chapter 6 concentrates on optical-layer protection and restoration. The need for optical layer survivability is explained, and the fundamental concepts of protection and restoration for optical layer survivability are introduced. A variety of basic protection and restoration schemes are presented, and both the survivable network design for static traffic and the survivable routing for dynamic traffic are discussed. For static traffic, the survivable network design can be formulated as a MILP problem. Two examples of ILP formulations are given, which are based on dedicated path protection and shared path protection, respectively. For dynamic traffic, various survivable routing algorithms are discussed. Moreover, this chapter explores dynamic restoration and presents several dynamic restoration protocols for surviving single link failures in the network.

Chapter 7 covers the major issues related to IP over WDM networks. Various IP over WDM layered models, service models, and interconnection models are introduced. The MPLmS and G-MPLS control planes for IP over WDM are described. Moreover, network survivability in IP over WDM networks is discussed.

Chapter 8 concludes the book with a brief discussion on future trends in optical WDM networks.

The appendices cover some basics of graph theory, *Dijkstra*'s algorithm, and acronyms.

Problems

1.1 How large a bandwidth can a single single-mode fiber potentially provide? How many orders of magnitude is this bandwidth higher than the currently achievable electronic processing speed?

1.2 What are the main significant characteristics of optical fibers?

1.3 Why is it difficult to make use of the total bandwidth of an optical fiber by using a single wavelength?

1.4 What is the bandwidth of OC-48 and OC-192?

1.5 Where are the two low-attenuation areas in the transmission spectrum of an optical fiber? How low is the attenuation in these two areas?

1.6 What are the basic components of a WDM transmission system? Describe their basic functions and characteristics.

1.7 Consider a usable bandwidth of 500 Gbps. If the frequency spacing is 2.5 GHz, how many OC-48 channels are available? If the frequency spacing is 20 GHz, how many OC-192 channels are available?

1.8 Why does a broadcast-and-select WDM network need a large number of wavelengths? Why is a broadcast-and-select WDM network more suitable for deployment in LANs and MANs?

1.9 What is the wavelength-continuity constraint? Give an example to explain how this constraint affects the network performance in terms of wavelength utilization and blocking probability.

1.10 What does data and protocol transparency refer to? Why can the optical layer support data and protocol transparency?

References

[1] Rajiv Ramaswami and Kumar N. Sivarajan, *Optical Networks—A Practical Perspective*, Second Edition, Morgan Kaufmann Publishers, San Francisco, 2002.

[2] B. Mukherjee, *Optical Communication Networks*, McGraw Hill, New York, 1997.

[3] Paul E. Green, *Fiber-Optic Networks*, Prentice Hall, Englewood Cliffs, New Jersey, 1993.

[4] Paul Green, "Progress in optical networking," *IEEE Communications Magazine*, vol. 39, no. 1, Jan. 2001, pp. 54–61.

[5] Dirceu Cavendish, "Evolution of optical transport technologies: from SONET/SDH to WDM," *IEEE Communications Magazine*, vol. 38, no. 6, Jun. 2000, pp. 164–172.

[6] Jaafar M. H. Elmirghani and Hussein T. Mouftah, "Technologies and architectures for scalable dynamic dense WDM networks," *IEEE Communications Magazine*, vol. 38, no. 2, Feb. 2000, pp. 58–66.

[7] Malathi Veeraraghavan et al., "Architectures and protocols that enable new applications on optical networks," *IEEE Communications Magazine*, vol. 39, no. 3, Mar. 2001, pp. 118–127.

[8] Nasir Ghani, Sudhir Dixit, and Ti-Shiang Wang, "On IP-over-WDM integration," *IEEE Communications Magazine*, vol. 38, no. 3, Mar. 2000, pp. 72–84.

[9] Chadi Assi, "Optical networking and real-time provisioning: An integrated vision for the next-generation Internet," *IEEE Network*, vol. 15, no. 4, Jul./Aug. 2001, pp. 36–45.

[10] Paul Bonenfant and Antonio Rodriguez-Moral, "Optical data networking," *IEEE Communications Magazine*, vol. 38, no. 3, Mar. 2000, pp. 63–70.

[11] Muriel Medard and Steven Lumetta, "Architecture issues for robust optical access," *IEEE Communications Magazine*, vol. 39, no. 7, Jul. 2001, pp. 116–122.

[12] Viktoria Elek, Andrea Fumagalli, and Gosse Wedzinga, "Photonic slot routing: A cost-effective approach to designing all-optical access and metro networks," *IEEE Communications Magazine*, vol. 39, no. 11, Nov. 2001, pp. 164–172.

[13] Mark Kuznetsov et al., "A next-generation optical regional access network," *IEEE Communications Magazine*, vol. 38, no. 1, Jan. 2000, pp. 66–72.

[14] John M. Senior, Michael R. Handley, and Mark S. Leeson, "Developments in wavelength division multiple access networking," *IEEE Communications Magazine*, vol. 36, no. 12, Dec. 1998, pp. 28–36.

[15] Leonid G. Kazovsky, Thomas Fong, and Tad Hofmeister, "Optical local area network technologies," *IEEE Communications Magazine*, vol. 32, no. 12, Dec. 1994, pp. 50–54.

[16] C. Siva Ram Murthy and Mohan Gurusamy, *WDM Optical Networks: Concepts, Design, and Algorithms*, Prentice Hall PTR, Upper Saddle River, New Jersey, 2002.

[17] Hui Zang, Jason P. Jue, and Biswanath Mukherjee, "A review of routing and wavelength assignment approaches for wavelength-routed optical WDM networks," *SPIE Optical Networks Magazine*, vol. 1, no. 1, Jan. 2000, pp. 47–60.

[18] R. Dutta and G. N. Rouskas, "A survey of virtual topology design algorithms for wavelength routed optical networks," *SPIE Optical Networks Magazine*, vol. 1, no. 1, Jan. 2000, pp. 73–89.

[19] Jun Zheng and Hussein T. Mouftah, "Distributed lightpath control for wavelength-routed WDM networks," *The Handbook of Optical Communication Networks*, CRC Press LLC, Chapter 15, Boca Raton, Florida, Apr. 2003, pp. 273–286.

[20] Hui Zang et al., "Dynamic lightpath establishment in wavelength-routed WDM networks," *IEEE Communications Magazine*, vol. 39, no. 9, Sep. 2001, pp. 100–108.

[21] Ornan Gerstel and Rajiv Ramaswami, "Optical layer survivability: A services perspective," *IEEE Communications Magazine*, vol. 38, no. 3, Mar. 2000, pp. 104–113.

[22] Ornan Gerstel and Rajiv Ramaswami, "Optical layer survivability: An implementation perspective," *IEEE Journal on Selected Areas in Communications*, vol. 18, no. 10, Oct. 2000, pp. 1885–1899.

[23] Dongyun Zhou and Suresh Subramaniam, "Survivability in optical networks," *IEEE Network*, vol. 14, no. 6, Nov./Dec. 2000, pp. 16–23.

[24] S. Ramamurthy and Biswanath Mukherjee, "Survivable WDM mesh networks, part I—Protection," *Proceedings of IEEE INFOCOM'99*, vol. 2, New York, pp. 744–751.

[25] S. Ramamurthy and Biswanath Mukherjee, "Survivable WDM mesh networks, part II—Restoration," *Proceedings of ICC'99*, vol. 3, Vancouver, Canada, pp. 2023–2030.

[26] Antonio R. Moral, Paul Bonenfant, and Murali Krishnaswamy, "The optical Internet: Architectures and protocols for the global infrastructure of tomorrow," *IEEE Communications Magazine*, vol. 39, no. 7, Jul. 2001, pp. 152–159.

[27] Marco Listanti, Vincenzo Eramo, and Roberto Sabella, "Architectural and technological issues for future optical Internet networks," *IEEE Communications Magazine*, vol. 38, no. 9, Sep. 2000, pp. 82–92.

[28] Bala Rajagopalan et al., "IP over optical networks: architectural aspects," *IEEE Communications Magazine*, vol. 38, no. 9, Sep. 2000, pp. 94–102.

Chapter 2

Fundamentals of WDM Network Devices

2.1 Introduction

WDM technology is evolving from a point-to-point transmission technology to a networking technology. The deployment of wavelength-routed WDM networks depends highly on advances in enabling technologies for optical devices. To better understand the subsequent chapters of this book, it is necessary to have some fundamental knowledge about the major optical devices that are used in wavelength-routed WDM networks before we move ahead. This chapter is intended to give a brief introduction to such devices, including optical fibers, couplers, amplifiers, transmitters and receivers, multiplexers and demultiplexers, optical add/drop multiplexers (OADMs), optical cross-connects (OXCs), and wavelength converters. To help those readers who have no background in fiber optics, we will try to minimize the optics involved and focus on the basic functions and characteristics. For a more detailed discussion of the principles behind, the readers are referred to [1–2].

2.2 Optical Fibers

Optical fiber is an excellent physical medium for high-speed transmission. It can provide extremely low-attenuation transmission over a huge frequency range and thus has a number of advantages over traditional transmission media such as copper and air. As mentioned in Section 1.2, a single single-mode fiber can provide a transmission bandwidth of about 50 Tbps, which is many orders of magnitude more than the bandwidth available in copper or any other transmission medium. In particular, the attenuation in the 1300 nm and 1500 nm areas is less than 0.5 dB/km, which allows an optical signal to propagate in an optical fiber over a long distance without any amplification or regeneration. It is because of these facts that optical fibers have been used so widely in today's communication systems and networks.

An optical fiber consists of a fine cylindrical glass core surrounded by a glass cladding. The cladding is usually protected by a thin plastic jacket. The cross section of an optical fiber is illustrated in Figure 2.1. Both the core and the cladding have an index of refraction, and the index of refraction of the core is made slightly higher than that of the cladding. Note that the index of refraction of a substance is defined as the ratio of the light speed in vacuum to the light speed in the substance.

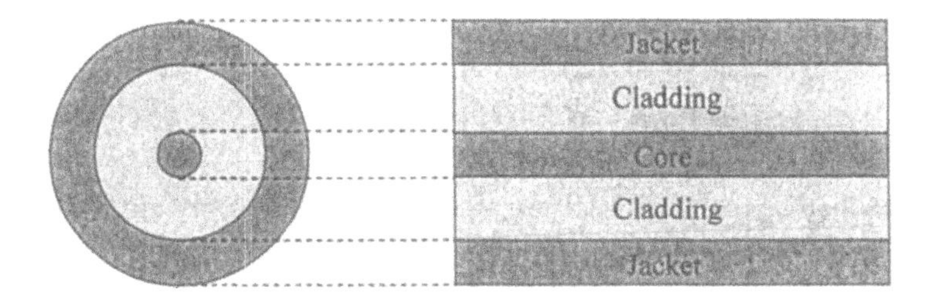

Figure 2.1 Cross and longitudinal sections of an optical fiber.

From the viewpoint of geometric optics, light consists of a number of light rays propagating in straight directions in a medium, which are reflected and refracted at the interface between two different media. For optical fibers, the ratio of the index of refraction in the cladding to that in the core defines a so-called critical angle (i.e., θ_c), as shown in Figure 2.2. If a light ray from the core is incident on the interface of the core and the cladding at an angle less than the critical angle, all energy of the ray will be completely reflected back into the core. This phenomenon is called total internal reflection, and a light ray that undergoes a total internal reflection is called a guided ray. Otherwise, the ray will be partially refracted to the cladding and is thus called an unguided ray. It is because of the total internal reflection phenomenon that light can propagate in optical fibers with very low attenuation over a long distance.

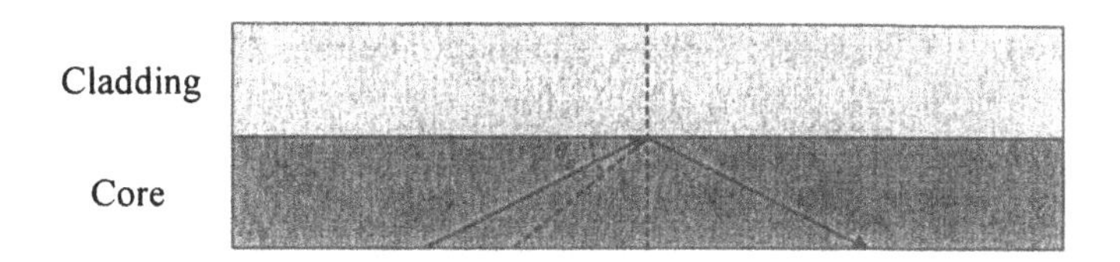

Figure 2.2 Reflection in optical fiber.

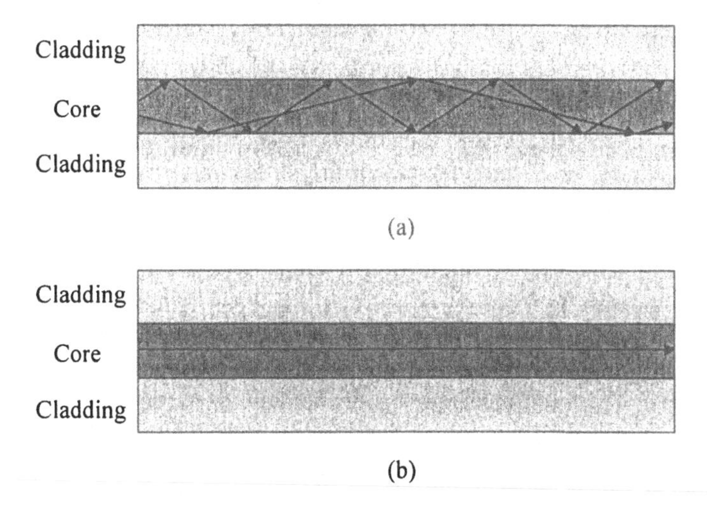

Figure 2.3 Propagation of light rays: (a) in multimode fibers; (b) in single-mode fibers.

An optical fiber with many different guided rays propagating inside the core is called a multimode fiber. This type of fiber generally has a core radius that is much larger than the operating wavelengths. In a multimode fiber, different guided rays have different incident angles on the interface of the core and the cladding, and thus take different propagating paths at different speeds. Each of the guided rays is said to propagate in a different mode, which causes a phenomenon called intermodal dispersion. As we will see later, intermodal dispersion largely limits the maximum bit rate and the transmission distance over the fiber. To eliminate intermodal dispersion, the core radius of an optical fiber must be very small so that only one mode exists in the fiber. This type of fiber is called a single-mode fiber. In a single-mode fiber, there is only one mode in which a light ray can propagate and the light propagates in a straight direction along the fiber. As a result, data can be transmitted in a single-mode fiber at a few gigabits per second over hundreds of kilometers without any amplification. Typically, the core radius of a multimode fiber is about 25μm whereas that of a single-mode fiber is about 4μm [2]. Figure 2.3 illustrates the propagation of light rays in multimode fibers and single-mode fibers.

An optical signal suffers from a variety of transmission impairments as it propagates in fibers. The most significant transmission impairments include attenuation, dispersion, and nonlinearity, which would impose limits on the transmission distance, the transmission bit rate, and the transmission quality of an optical signal and thereby degrade the performance of an optical transmission system.

Attenuation in fiber

Attenuation in an optical fiber leads to the loss of signal power as an optical signal propagates over some distance through the fiber. Attenuation loss is primarily caused by the material absorption and Rayleigh scattering in optical fibers [2]. Material absorption includes the resonance absorption in silica as well as that by the impurities in silica, which is a major loss factor. However, the loss by material absorption has been reduced to a negligible level on the wavelengths of interest today. As a result, Rayleigh scattering becomes the dominant factor for attenuation loss, especially for the short wavelengths. Rayleigh scattering occurs because of the small fluctuations in the density of the medium, which result in the light being scattered and thus attenuate the propagating light wave. On the other hand, the attenuation loss of an optical signal depends on the wavelength on which it is transmitted. As shown in Figure 1.1, fibers have three relatively flat low-loss areas centered at 850 nm, 1300 nm, and 1550 nm. Early optical transmission systems operated in the 850 nm area, which has an attenuation of about 0.8 dB/km. The minimum attenuation occurs in the 1550 nm area and is approximately 0.2 dB/km. The 1300 nm area has an attenuation of about 0.5 dB/km. Most optical transmission systems today operate in the 1550 nm area.

Attenuation loss imposes a limit on the transmission distance of an optical signal. To overcome this limit, electronic regenerators can be used at appropriate locations in optical transmission systems to restore the attenuated signal for further transmission. In particular, with the advent of the erbium-doped fiber amplifier (EDFA), this limit has been reduced to a great degree (see Section 2.4).

Dispersion in fiber

Dispersion refers to the phenomenon in which a signal pulse spreads out as it propagates along a fiber. When a signal pulse spreads out to the extent that it overlaps its neighboring pulses, intersymbol interference occurs, which would largely increase the bit error rate (BER) at the receiver. This phenomenon imposes a limit on the minimum interval between two consecutive pulses and hence on the transmission rate of an optical pulse signal. There are two basic dispersive effects in a fiber: intermodal dispersion and chromatic dispersion. Intermodal dispersion occurs only in multi-mode fibers. Because a signal pulse propagates in different modes and each mode propagates at a different speed, different modes would arrive at the receiver with different propagation delays and thus result in the pulse spreading out. Obviously, intermodal dispersion is not a problem with single-mode fibers. Chromatic dispersion occurs because of the frequency dependence of the propagation speed. There are two kinds of chromatic dispersion: material dispersion and waveguide dispersion. Material dispersion is caused by the different refractive indexes in silica for different wavelengths. Waveguide dispersion is caused by the different characteristics of the fiber core and cladding for different wavelengths.

Nonlinearity in fiber

Nonlinear effects in fiber may potentially lead to signal attenuation, distortion, and interference, and thereby have a significant impact on the performance of WDM transmission systems. There are a variety of nonlinear effects, including self-phase modulation (SPM), four-wave mixing (FWM), cross-phase modulation (XPM), stimulated Raman scattering (SRS), and stimulated Brillouin scattering (SBS). These effects impose limits on the spacing between two adjacent optical channels, the maximum power of a single signal, and the maximum transmission rate. It has been shown that a WDM transmission system with a channel spacing of 10 GHz and a transmitting power of 0.1 mW per channel can provide a maximum of about 100 channels in the 1550 low-attenuation area [3]. For further details on nonlinear effects in fiber, the readers are referred to [4].

2.3 Couplers

A coupler is an optical device that is used to combine or split optical signals in an optical network. Typically, a 2×2 coupler consists of two input ports and two output pots, as shown in Figure 2.4. One implementation of a 2×2 coupler is to fuse two fibers together. In such a coupler, one fraction of the signal power from input port 1 is coupled to output port 1 while the other fraction is coupled to output port 2. If the fraction to output port 1 is α, the fraction to output port 2 is $1-\alpha$, where α is referred to as the coupling coefficient. Likewise, for input port 2, a fraction $1-\alpha$ of the signal power is coupled to output port 1 while a fraction α is coupled to output port 2.

Figure 2.4 A 2×2 coupler.

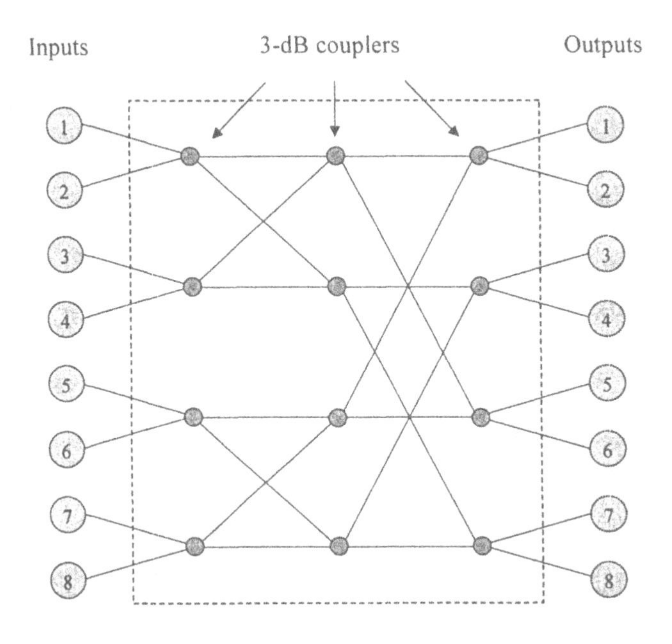

Figure 2.5 The structure of an 8×8 passive star coupler.

The coupling coefficient α of a coupler can be designed at different values for a variety of purposes. For example, a coupler can be used as a splitter that divides an optical signal from one fiber to two fibers. In this case, half the power from each input port is coupled to each output port, i.e. $\alpha=0.5$, and such a coupler is called a 3-dB coupler. A typical application of a 3-dB coupler is to build an $N \times N$ passive star coupler, which is a multiport device that divides the power from each input port equally among all of its output ports. Figure 2.5 shows the structure of an 8×8 passive star coupler. A coupler can also be used to tap off a small portion of the signal power in a fiber. In this case, the coupler is usually designed with a value of α close to 1, typically 0.90–0.95, and is used for monitoring or other purposes. In addition, couplers can be used as a building block for other optical devices, such as optical filters, multiplexers and demultiplexers, and optical switches, which will be introduced in the subsequent sections.

On the other hand, a coupler can be made either wavelength selective or wavelength independent. For a wavelength-selective coupler, its coupling coefficient depends on the wavelength of an optical signal. It is widely used to combine two signals on different wavelengths into a single fiber without any loss. In contrast, a wavelength-independent coupler has the same coupling coefficient over a wide range of wavelengths.

2.4 Optical Amplifiers

An optical amplifier is used to amplify the power of an optical signal in an optical transmission system. Before the advent of optical amplifiers, this was implemented by using an electronic regenerator. An electronic regenerator converts the optical signal to an electronic signal, processes it, and then converts the electronic signal back into an optical signal for retransmission. There are three types of regenerators: 3R regenerator (i.e., regeneration, reshaping, and retiming), 2R regenerator (i.e., regeneration and reshaping), and 1R (regeneration) regenerator. Both 3R and 2R regenerators are electronic regenerators, which have been used widely in today's optical transmission systems and networks. 1R regenerators actually refer to optical amplifiers, which perform signal amplification optically without reshaping and retiming operations.

Optical amplifiers have several advantages over electronic regenerators. Optical amplifiers are insensitive to the bit rates and modulation formats of optical signals and can thus provide data transparency to the bit rates and modulation formats. As a result, an optical transmission system using optical amplifiers is more easily upgraded to a higher bit rate or a different modulation format without replacing the amplifiers. In contrast, regenerators are sensitive to bit rates and modulation formats. With regenerations, a system upgrade may require all the regenerators to be replaced. Moreover, optical amplifiers have fairly large gain bandwidths. A single amplifier can simultaneously amplify several optical signals on different wavelengths. In contrast, a regenerator can amplify only one wavelength. A WDM transmission system must use a multiplexer and a demultiplexer to separate each wavelength before amplifying each signal electronically.

An optical amplifier can typically serve as a power amplifier, a line amplifier, and a preamplifier at different points in an optical transmission system for different purposes. As a power amplifier, it is used to raise the power level of an optical signal at the output of a transmitter before it is transmitted to a transmission link. The purpose is to increase the transmission distance before any amplification is needed. As a line amplifier, it is placed at appropriate points along the transmission link and is used to compensate for the power loss of a signal as the signal propagates over a long distance. As a preamplifier, it is used to raise the signal level at the input of a receiver in order to increase the sensitivity of the receiver. In addition, an optical amplifier can also be used at other points in an optical network, such as in optical switches.

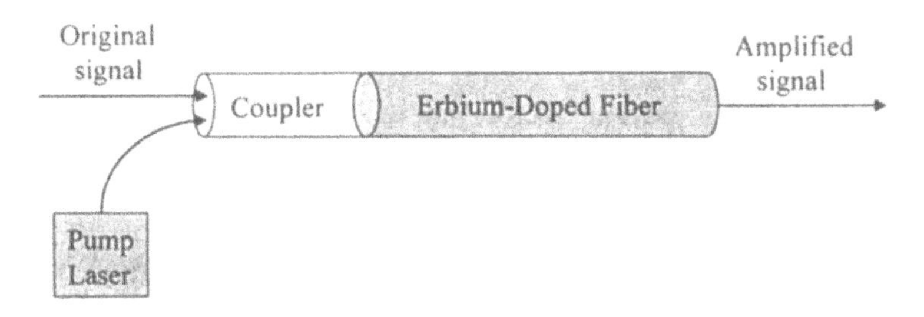

Figure 2.6 An erbium-doped fiber amplifier.

There are typically two basic types of optical amplifiers: rare earth-doped fiber amplifiers and semiconductor optical amplifiers (SOAs). A rare earth-doped fiber amplifier consists of a length of single-mode fiber, typically on the order of tens of meters long, with the core doped with a rare earth element, as shown in Figure 2.6. The most commonly used rare earth element is erbium. An erbium-doped fiber amplifier (EDFA) can amplify optical signals over a range of wavelengths between 1525 nm and 1560 nm. The erbium-doped fiber is pumped by a pump signal from a laser, which supplies power to the doped fiber at a wavelength of 980 nm or 1480 nm. The pump is typically coupled into the doped fiber through a wavelength-selective coupler. It is the power of the pump signal that is transferred to the signal power as the pump signal propagates along the doped fiber. It has been shown that the 980-nm pump signal can provide a gain of about 10 dB/mW, whereas the 1480-nm pump signal can provide a gain of about 5 dB/mW. Typically, EDFAs can achieve a gain of about 25 dB. However, they have been shown to be capable of achieving a gain of up to 51 dB with the maximum gain limited by internal Rayleigh backscattering [5]. The most significant characteristic of EDFAs is that they can provide a fairly wide bandwidth of about 35 nm.

Semiconductor optical amplifiers (SOAs) use semiconductors to amplify optical signals. This type of amplifier can provide a bandwidth of the order of 100 nm, which is much larger than the bandwidth EDFAs can provide. In practice, signals in the 1.3-nm and 1.5-nm areas can even be simultaneously amplified with SOAs. Although SOAs have this advantage, they are less preferred than EDFAs for use in WDM systems for several reasons [2]. The main reason is that SOAs introduce severe cross talk when they are used in WDM systems. Moreover, compared with EDFAs, SOAs provide lower power gains and result in higher coupling loss and polarization-dependent loss. For more details on SOAs and EDFAs, the readers are referred to [2] and [6].

2.5 Transmitters

An optical transmitter is a light source that generates an optical signal operating at a particular wavelength, which can be modulated by an electrical signal. A typical example of modulation is the so-called on-off keying (OOK), in which a light source is turned on or off under the control of a sequence of binary bits.

Optical transmitters can be typically classified into two categories: lasers (i.e., light amplification by stimulated emission of radiation) and LEDs (i.e., light-emitting diodes), where lasers can be further divided into semiconductor lasers and fiber lasers. However, the most widely used light sources in optical transmission systems today are semiconductor diode lasers. A semiconductor diode laser is basically a device for converting electronic energy to monochromatic light. It has several advantages over fiber lasers, such as compact size, ease in integration, no pumping, suitability for large-scale integration, and high efficiency in energy conversion. Almost all optical transmission systems today use semiconductor lasers as their light sources. However, lasers are expensive and thus may not be cost effective for use in networks at low bit rates over short distances. In contrast, LEDs are inexpensive and are thus a cost-effective alternative choice.

On the other hand, lasers can also be classified into fixed-tuned lasers and tunable lasers. A fixed-tuned laser is a device that operates at a fixed wavelength. A typical application of fixed-tuned lasers is a laser array. A laser array consists of a set of fixed-tuned lasers, with each laser operating at a different wavelength. An advantage of the laser array is the simultaneous transmission of multiple signals on different wavelengths. However, a laser array is not flexible because the number of wavelengths available in it is fixed and limited. In contrast, a tunable laser is more flexible because it can tune its operating wavelength flexibly. For more details on different types of transmitters and lasers, the readers are referred to [1] and [7–8].

2.6 Receivers

An optical receiver is a device that converts a modulated optical signal received at its input port into an electronic data signal at its output port. It mainly consists of four basic components: a photodetector, a preamplifier, a front-end amplifier, and a recovery circuit, as shown in Figure 2.7. The photodetector generates an electronic current called a photocurrent proportional to the input optical power. The front-end amplifier increases the power of the photocurrent to a level high enough for further electronic processing. The preamplifier raises the signal level at the input of the receiver to increase the sensitivity of the receiver. The sensitivity of a receiver is the average optical power required to achieve an acceptable bit error rate at a particular bit rate. The modulated optical signal is transmitted over the optical fiber. It undergoes a variety of impairments, such as attenuation and dispersion, and has noise added to it from optical amplifiers during the transmission. To recover the transmitted data at the receiver, the recovery circuit performs clock recovery, sampling, and threshold detection to extract the digital bits. The photodetectors used in optical transmission systems are semiconductor photodiodes, which may use different detection technologies, such as direct detection and coherent detection [9]. For more details on different types of photodetectors, the readers are referred to [1], [7], and [9].

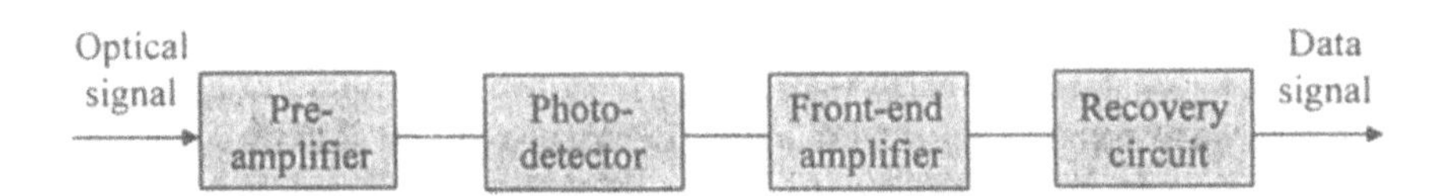

Figure 2.7 Block diagram of a receiver.

2.7 Optical Add/Drop Multiplexers

In wavelength-routed WDM networks, it is often necessary to drop some traffic at an intermediate node on an end-to-end path. An optical add/drop multiplexer (OADM), also called wavelength add/drop multiplexer, is a device for this purpose. The function of an OADM is to add and drop some wavelengths locally as necessary while at the same time bypassing the other wavelengths through the device directly. An OADM typically consists of a demultiplexer, 2×2 switches with one switch per wavelength, and a multiplexer, as shown in Figure 2.8. Each 2×2 switch has an electronic controller to control its configuration. If a 2×2 switch is configured into a bar state under electronic control, the signal at the corresponding wavelength is bypassed through the switch and the OADM as well. If a switch is in a cross state, the signal on the corresponding wavelength is dropped locally and another signal on the same wavelength can be added.

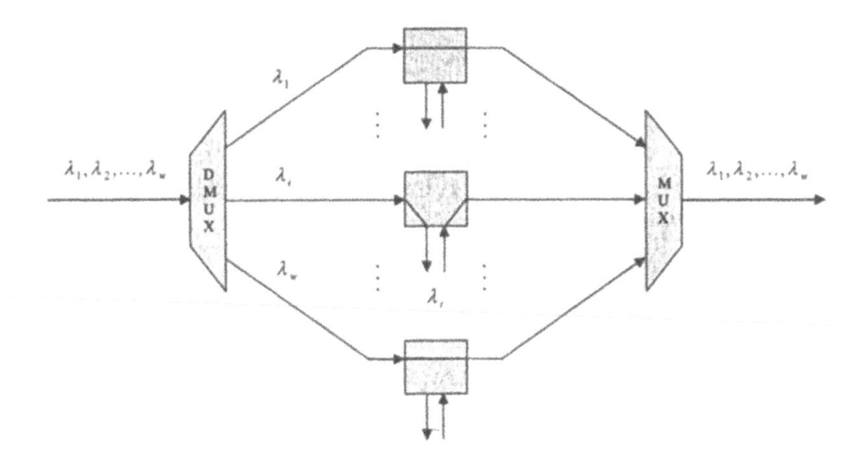

Figure 2.8 An optical add/drop multiplexer.

2.8 Optical Cross-Connects (OXCs)

An optical cross-connect (OXC) is also referred to as a wavelength cross-connect (WXC) or a wavelength router. It is an optical device that can route an input wavelength on an input port to an output port. OXCs can be classified into two basic categories: nonreconfigurable OXCs and reconfigurable OXCs. For a nonreconfigurable OXC, its configuration is fixed without changing over time. A wavelength on an input port can only be routed to a fixed output port. Figure 2.9 illustrates a 4×4 nonreconfigurable WXC, which consists of four multiplexers and four demultiplexers. There are four optical channels on each input link and each output link. The demultiplexers separate each of the optical channels on an input port, whereas the multiplexers recombine the optical channels from different input ports onto a single output port. Between the multiplexers and demultiplexers, there are fixed direct connections from each output of the demultiplexers to each input of multiplexers. The interconnection pattern is determined during the design of the OXC.

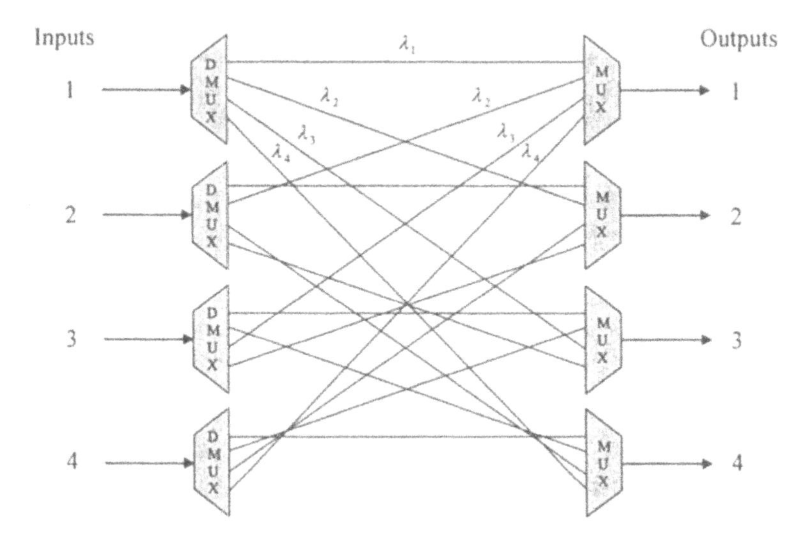

Figure 2.9 A 4×4 nonreconfigurable OXC.

In contrast, a reconfigurable OXC can be reconfigured under electronic control. It typically consists of demultiplexers, optical switches, and multiplexers. An $N \times N$ reconfigurable OXC is shown in Figure 2.10, in which there are W wavelengths or optical channels on each input fiber and each output fiber. Similar to a nonreconfigurable OXC, the demultiplexers separate each of the optical channels on an input port, whereas the multiplexers recombine the optical channels from different input ports to a single output port. Unlike a nonreconfigurable OXC, the outputs of the demultiplexers are directed to an array of W $N \times N$ optical switches between the demultiplexers and the multiplexers. All optical channels with the same wavelength are directed to the same switch. The outputs of the switches are then directed to the inputs of the multiplexers. Each switch uses an optical space-division switch to route an optical channel from an input port to any output port under electronic control.

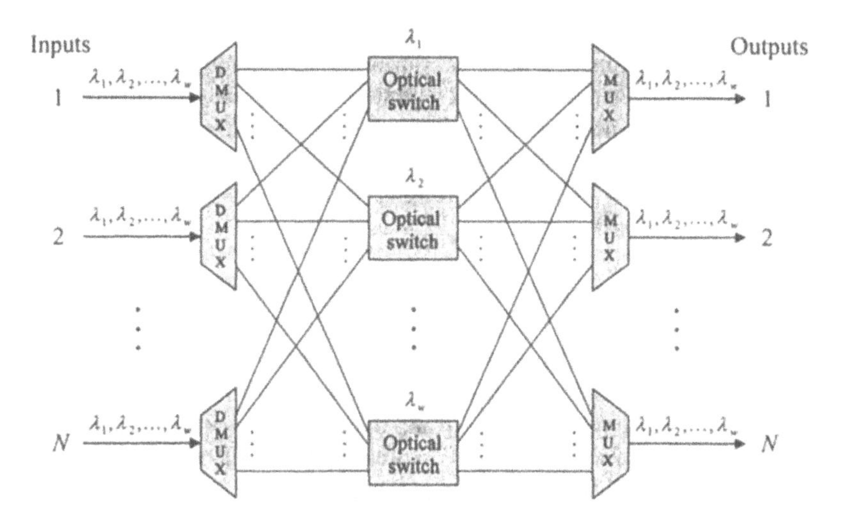

Figure 2.10 An $N \times N$ reconfigurable OXC.

An optical space-division switch can be made out of 2×2 cross-point elements [10]. A 2×2 cross-point element can route an optical signal from one input port to one output port. It may have two states: the bar state and the cross state, as shown in Figure 2.11. In the bar state, a signal from the upper input port is routed to the upper output port and the signal from the lower input port is routed to the lower output port. In the cross state, the signal from the upper input port is routed to the lower output port and the signal from the lower input port is routed to the upper output port. The switch states can be reconfigured under electronic control. Note that optical cross-point elements are usually wavelength insensitive.

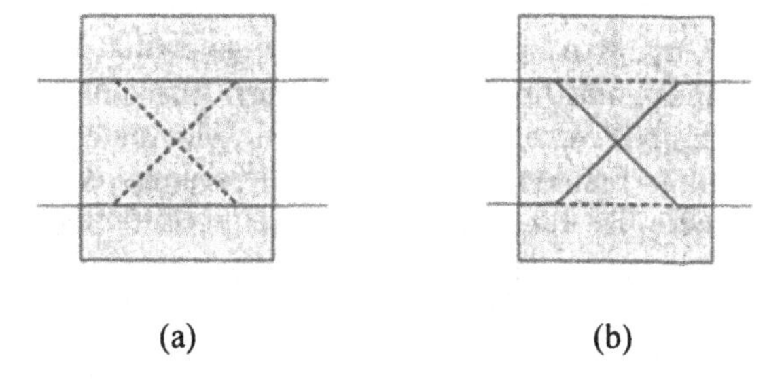

Figure 2.11 2×2 cross-point elements: (a) bar state; (b) cross state.

OXCs provide a high level of flexibility in terms of wavelength routing and switching capabilities in the optical domain. With OXCs, optical signals can be routed and switched through a routing node without undergoing any O/E/O conversion, which provides a high level of data transparency. The transparent routing and switching allows for OXCs to be independent of the bit rates and protocol formats of optical signals. It is with the advent of OXCs that all-optical WDM networking becomes possible.

2.9 Wavelength-Convertible Cross-Connects

OXCs provide a high degree of flexibility in terms of wavelength routing and switching. However, this type of device has no wavelength conversion capability so that a wavelength on an input port can only be routed and switched to the same wavelength on an output port. This imposes some limit on the routing capability of OXCs. To improve the routing capability, OXCs can use wavelength converters to provide a higher level of flexibility at a routing node. This can alleviate and even eliminate the wavelength-continuity constraint and thus improve the network performance in terms of wavelength utilization and blocking probability.

A wavelength-convertible cross-connect, also called wavelength-convertible switch (WCS), is an optical device that is capable of routing a wavelength on an input port to a different wavelength on an output port. There are typically three different types of WCS architectures: dedicated, share-per-node, and share-per-link [11].

Dedicated WCS architecture

In the dedicated architecture, there is a dedicated wavelength converter (WC) for each output port or wavelength of the optical switch, as shown in Figure 2.12. The optical switch can be configured appropriately to switch an input wavelength to any of its output ports. In general, a WCS with N input links, N output links, and W wavelength on each fiber link has an $NW \times NW$ optical switch and NW wavelength converters. The optical signal received on each input link is first demultiplexed into W separate signals on different wavelengths, which are directed to the input ports of the optical switch. Each wavelength is then switched to one of the output ports of the optical switch, and the output wavelength may be changed into a different wavelength at the converter. Finally, all the wavelengths that are routed to a particular output link are multiplexed into a single optical signal in the multiplexer of that particular output link. Because each output port or wavelength of the optical switch has a dedicated wavelength converter, this architecture can provide nonblocking switching for all the wavelengths.

However, it requires many wavelength converters and these converters may not be used efficiently. A cost-effective solution is to use only a few converters and allow them to be shared among all the wavelengths.

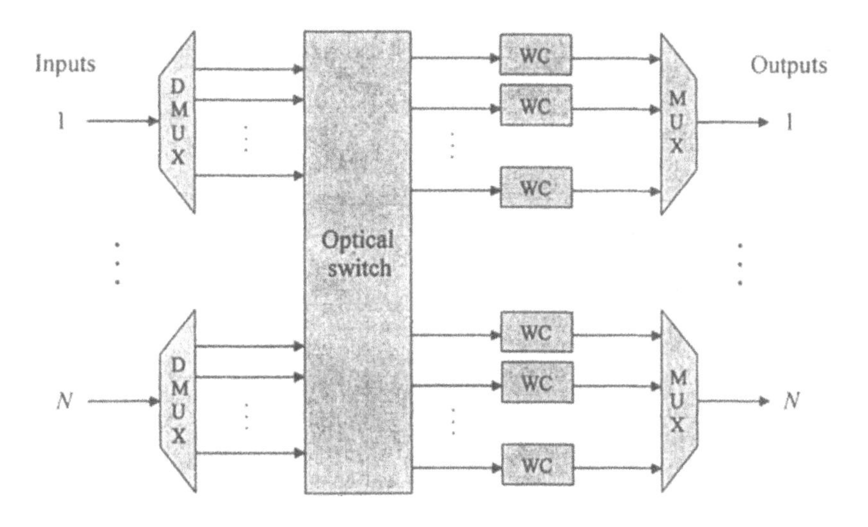

Figure 2.12 Architecture of dedicated WCS.

Share-per-node WCS architecture

The share-per-node architecture uses a couple of optical switches and a wavelength converter bank (WCB), as shown in Figure 2.13. The converter bank is a collection of wavelength converters that is shared among all the wavelengths on the input links and can be accessed by any wavelength on an input link by appropriately configuring the first optical switch. Only those wavelengths that require wavelength conversion are directed to the converter bank. Both optical switches can be configured appropriately to switch an input wavelength to any of its output ports. For a WCS with N input links, N output links, and W wavelengths on each fiber link, the first optical switch is a $NW \times (NW+C)$ nonblocking switch and the second switch is a $C \times NC$ nonblocking switch, where C is the number of converters in the converter bank. The optical signal received on each input link is first demultiplexed into W separate signals on different wavelengths, which are directed to the first optical switch. If a wavelength does not require wavelength conversion, it is switched to an output port of the first optical switch, which is directed to the multiplexer of the corresponding output link. Otherwise, the wavelength is switched to one of the output ports that are directed to the converter bank.

The converted wavelength is directed to the second switch, where it is switched to an output port that is directed to the multiplexer of the corresponding output link. All the wavelengths that are routed to a particular output link are multiplexed into a single optical signal in the multiplexer of that particular output link. Because all wavelength converters in the converter bank are shared among all the wavelengths, this architecture provides a cost-effective solution.

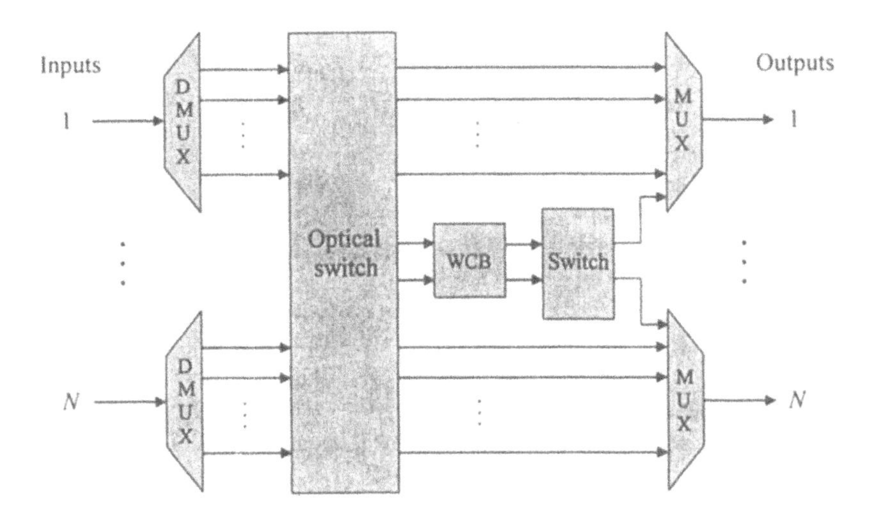

Figure 2.13 Architecture of share-per-node WCS.

Share-per-link WCS architecture

In the share-per-link architecture, each output link has a dedicated wavelength converter bank associated with it, as shown in Figure 2.14. Each dedicated converter bank can be shared and accessed only by those wavelengths that require wavelength conversion and are routed to the particular output link associated with it. The optical switch can be appropriately configured to switch an input wavelength to any of its output ports. For a WCS with N input links, N output links, and W wavelengths on each fiber link, the optical switch is a $NW \times (NW+NC)$ nonblocking switch, where C is the number of converters in each converter bank. The optical signal received on each input link is first demultiplexed into W separate signals on different wavelengths, which are directed to the optical switch. If a wavelength requires wavelength conversion, it is switched to an output port that is directed to the converter bank associated with the corresponding output link and the converted wavelength is directed to the multiplexer. Otherwise, it is switched to an output port that is directed to the multiplexer of the corresponding output link. All the wavelengths that are routed to a particular output link are multiplexed into a single optical signal in the multiplexer of that particular output link. Because a converter bank is only dedicated to a particular output link, the efficiency of converter sharing is less than that of the share-per-node architecture.

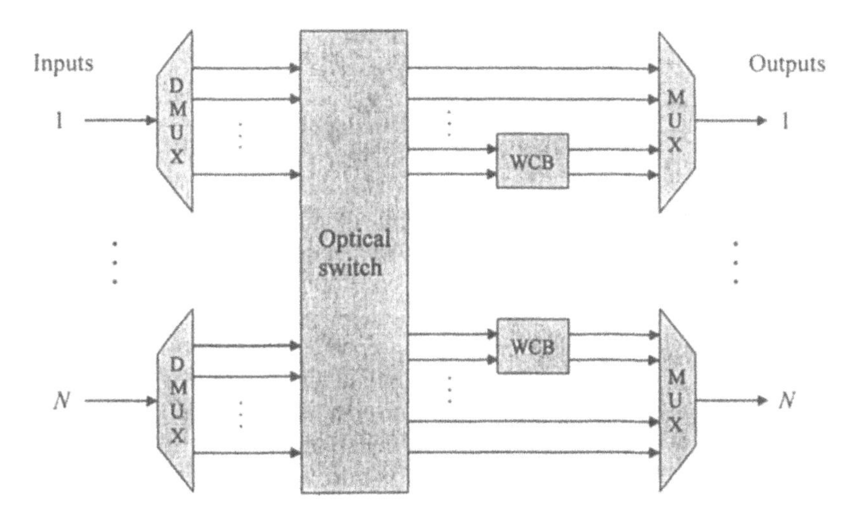

Figure 2.14 Architecture of share-per-link WCS.

Among the three different WCS architectures, the dedicated architecture can achieve the highest performance whereas the share-per-node architecture provides the lowest performance in terms of connection blocking probability. However, the share-per-node architecture has the highest efficiency in converter sharing and thus is the most cost-effective solution. The dedicated architecture does not allow any converter sharing and is therefore not cost effective. In addition, the implementation complexity of the dedicated architecture is the lowest whereas the share-per-node architecture has the highest complexity. The share-per-link architecture provides a trade-off solution in terms of performance, sharing efficiency, and implementation complexity.

2.10 Wavelength Converters

A wavelength converter is an optical device that is capable of converting one wavelength to another wavelength, as shown in Figure 2.15.

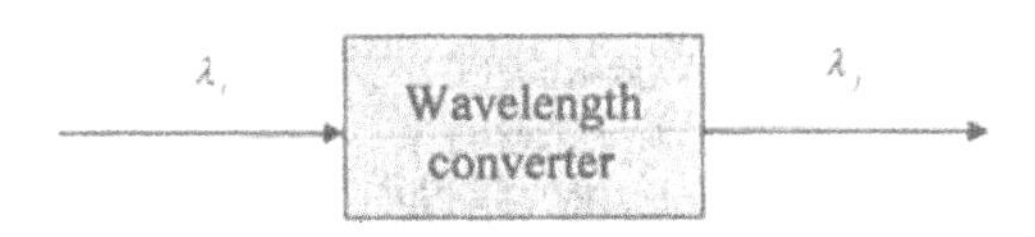

Figure 2.15 A wavelength converter.

A wavelength converter plays an important role in wavelength-routed WDM networks. The most significant benefit of using wavelength converters is eliminating the wavelength-continuity constraint [2] and improving the wavelength utilization in the network. The conversion capability of a wavelength converter is characterized by a conversion degree. A converter that is capable of converting one wavelength to any of W wavelengths is said to have a conversion degree of W. A converter is said to have a full degree of conversion if the conversion degree equals the number of wavelengths available on a fiber link. Otherwise, it is said to have a limited degree of conversion. In general, a wavelength converter is expected to have the following characteristics [12]:

- Transparency to bit rates and signal formats
- Different conversion capabilities
- Large conversion bandwidth
- Fast conversion time
- Insensitivity to input signal polarization
- Large signal-to-noise ratio
- Simple implementation

On the basis of the range of wavelengths that they can deal with at their input ports and output ports, wavelength converters can be classified into the following four types:

- Fixed-input and fixed-output converter: This type of converter always takes in a fixed-input wavelength and converts it into a fixed-output wavelength.

- Variable-input and fixed-output converter: This type of converter takes in a variable-input wavelength and always converts it into a fixed-output wavelength.
- Fixed-input and variable-output converter: This type of converter takes in a fixed-input wavelength and converts it into a variable-output wavelength.
- Variable-input and variable-output converter: This type of converter takes in a variable-input wavelength and converts it into a variable-output wavelength.

These types of wavelength converters can be used as a component of other optical devices, such as OXCs and OADMs, to provide different conversion capabilities at the network nodes. An OADM that uses fixed-input and fixed-output wavelength converters to provide different wavelength conversion capabilities is illustrated in Figure 2.16. Note that only the wavelengths that bypass the OADM are shown.

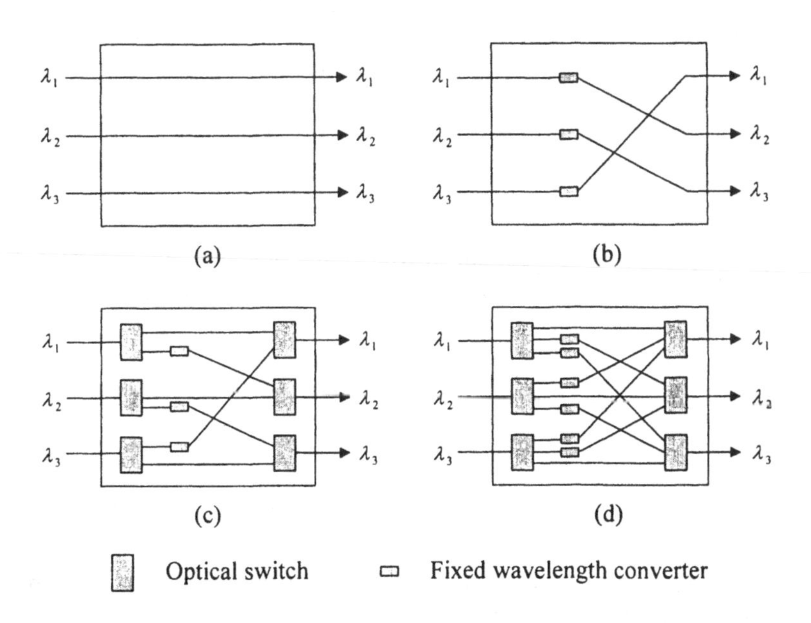

Figure 2.16 An OADM with different wavelength conversion capabilities: (a) no; (b) fixed; (c) limited; (d) full.

No wavelength conversion: A wavelength on an input port is always converted to the same wavelength on an output port, as shown in Figure 2.16(a).

Fixed wavelength conversion: A wavelength on an input port is always converted to another different wavelength on an output port, as shown in Figure 2.16(b).

Limited wavelength conversion: A wavelength on an input port can be converted to any of a set of wavelengths on an output port, as shown in Figure 2.16(c).

Full wavelength conversion: A wavelength on an input port can be converted to any wavelength on an output port, as shown in Figure 2.16(d).

In terms of the enabling technologies, wavelength converters can be broadly classified into two categories: opto-electronic wavelength converters and all-optical wavelength converters [2]. In an opto-electronic wavelength converter, an optical signal on a wavelength is first converted to the electronic form in a photodetector and the electronic signal is then used to modulate a laser at a different wavelength. This type of converter can operate at a bit rate up to 10 Gbps [13]. However, it has a higher implementation complexity and consumes much more power than all-optical converters. Moreover, the optical signal undergoes O/E and E/O conversion during the conversion process, which would largely reduce the degree of transparency. In an all-optical wavelength converter, an optical signal remains in the optical domain throughout the conversion process, which provides a high degree of transparency. All-optical conversion can be implemented by using coherent effects and cross-modulation [13]. For further details on different wavelength conversion technologies, the readers are referred to [1–2] and [13–14].

2.11 Summary

The deployment of wavelength-routed WDM networks largely depends on the advances in enabling technologies in fiber optics. Recent advances in optical devices, in particular, the advent of reconfigurable optical devices such as OADMs and OXCs, have made it possible to implement WDM networking and practically deploy wavelength-routed WDM networks. In this chapter, we have given a conceptual and functional introduction to the major optical devices that are used in wavelength-routed WDM networks, including optical fibers, couplers, amplifiers, transmitters and receivers, multiplexers and demultiplexers, optical add/drop multiplexers (OADMs), optical cross-connects (OXCs), and wavelength converters. Most of these devices are commercially available. Relatively speaking, all-optical wavelength converters are still immature devices. Although there is an increasing demand for such devices, their practical deployment still depends on significant cost reduction and performance improvement. The reconfigurable optical devices not only have had provided networking capabilities for optical networks but also have a great impact on the design of optical networks. As enabling technologies in optics continue to develop, optical devices with stronger networking capabilities are expected to become commercially available. Network designers can take advantage of such devices to improve network performance and provide better network service.

Problems

2.1 What is a critical angle or how is a critical angle defined?

2.2 What is the phenomenon of total internal reflection?

2.3 What are the major differences between a multimode fiber and a single-mode fiber?

2.4 What are the major transmission impairments in optical fibers? What effects do they have on an optical transmission system?

2.5 How many 3-dB couplers are needed to design a 8×8 star coupler? What about a 4×4 star coupler?

2.6 What advantages do optical amplifiers have over electronic regenerators?

2.7 What types of amplifiers can an optical amplifier typically serve as? Describe the main functions of these types of amplifiers.

2.8 What is an erbium-doped fiber amplifier (EDFA)? What is the most significant characteristic of EDFAs for optical networks?

2.9 Design a 4×4 share-per-node WCS. Suppose that there are four wavelengths on each fiber link and two wavelength converters shared among all wavelengths.

2.10 What are the main advantages of all-optical wavelength converters over opto-electronic converters?

References

[1] Paul E. Green, *Fiber-Optic Networks*, Prentice Hall, Englewood Cliffs, New Jersey, 1993.

[2] Rajiv Ramaswami and Kumar N. Sivarajan, *Optical Networks—A Practical Perspective*, Second Edition, Morgan Kaufmann Publishers, San Francisco, 2002.

[3] A. R. Chraplyvy, "Limits on lightwave communications imposed by optical-fiber nonlinearities," *IEEE/OSA Journal of Lightwave Technology*, vol. 8, no. 10, Oct. 1990, pp. 1548–1557.

[4] G. P. Agrawal, *Nonlinear Fiber Optics*, Academic Press, San Diego, California, 1989.

[5] S. L. Hansen, K. Dybdal, and L. C. Larsen, "Gain limit in erbium-doped fiber amplifiers due to internal Rayleigh backscattering," *IEEE Photonics Technology Letters*, vol. 4, no. 6, Jun. 1992, pp. 559–561.

[6] M. J. O'Mahony, "Optical Amplifiers," *Photonics in Switching*, vol. 1, Academic Press, San Diego, CA, 1993, pp. 147–167.

[7] C. A. Brackett, "Dense wavelength division multiplexing networks: principles and applications," *IEEE Journal on Selected Areas in Communications*, vol. 8, no. 6, Aug. 1990, pp. 948–964.

[8] T.-P. Lee and C.-E. Zah, "Wavelength-tunable and single-frequency lasers for photonic communications networks," *IEEE Communications Magazine*, vol. 27, no. 10, Oct. 1989, pp. 42–52.

[9] H. Kobrinski and K.-W. Cheung, "Wavelength-tunable optical filters: applications and technologies," *IEEE Communications Magazine*, vol. 27, no. 10, Oct. 1989, pp. 53–63.

[10] R. V. Schmidt and R. C. Alferness, "Directional coupler switches, modulators, and filters using alternating δβ techniques," *Photonic Switching*, IEEE Press, New York, 1990, pp. 71–80.

[11] K.-C. Lee and V. O. K. Li, "A wavelength-convertible optical network," *IEEE/OSA Journal of Lightwave Technology*, vol. 11, no. 5, May/Jun. 1993, pp. 962–970.

[12] T. Durhuus et al., "All-optical wavelength conversion by semiconductor optical amplifiers," *IEEE/OSA Journal of Lightwave Technology*, vol. 14, no. 6, Jun. 1996, pp. 942–954.

[13] S. J. B. Yoo et al., "Transparent wavelength conversion by difference frequency generation in AlGaAs waveguides," *Proceedings of Optical Fiber Communication (OFC'96)*, vol. 2, 1996, pp. 129–131.

[14] B. Mikkelsen "Wavelength conversion devices," *Proceedings of Optical Fiber Communication (OFC'96)*, vol. 2, 1996, pp. 121–122.

Chapter 3

Routing and Wavelength Assignment

3.1 Introduction

Routing and wavelength assignment (RWA) is one of the most important problems in wavelength-routed WDM networks. To establish a lightpath, the network must decide a physical route and assign an available wavelength on each link of the decided route, which involves routing and wavelength assignment. Because of the limitation in the number of wavelengths available on each fiber link as well as the wavelength-continuity constraint in the absence of wavelength converters, the network may not be able to accommodate all connection requests. As a result, a connection request is likely to be blocked because of the unavailability of wavelength resources. For this reason, it is desirable to use an efficient RWA algorithm to establish lightpaths for connection requests. RWA plays a very important role in achieving good network performance.

In this chapter, we discuss the RWA problem in wavelength-routed WDM networks. The RWA problem and related concepts are first introduced, and the objectives of the RWA problem for static and dynamic traffic are then described. The integer linear programming (ILP) formulations for solving the static RWA problems and a variety of RWA algorithms for solving the dynamic RWA problem are presented. Moreover, the RWA fairness problem is discussed and effective solutions are presented. We also discuss the wavelength rerouting problem, introduce the basic rerouting operations and schemes, and present a well-known wavelength rerouting algorithm. The focus of this chapter is primarily on wavelength-selective networks with the wavelength-continuity constraint unless otherwise stated.

3.2 RWA Problem

In wavelength-routed WDM networks, data traffic is transferred between network nodes through all-optical connections called lightpaths. Figure 3.1 illustrates a wavelength-routed WDM network in which three lightpaths have been established between different pairs of network nodes. As discussed in Section 1.3.2, a lightpath is a unidirectional all-optical connection between two network nodes, which may span multiple fiber links without undergoing any O/E and E/O conversion at each intermediate node. Two lightpaths cannot share the same wavelength on a common fiber link, which is referred to as the wavelength-distinct constraint. In the absence of wavelength conversion, a lightpath must use the same wavelength on all the fiber links it traverses, which is referred to as the wavelength-continuity constraint. This constraint is unique to WDM networks and would largely reduce wavelength utilization and thus degrade network performance. To show the effect of the wavelength-continuity constraint, let us examine Example 3.1.

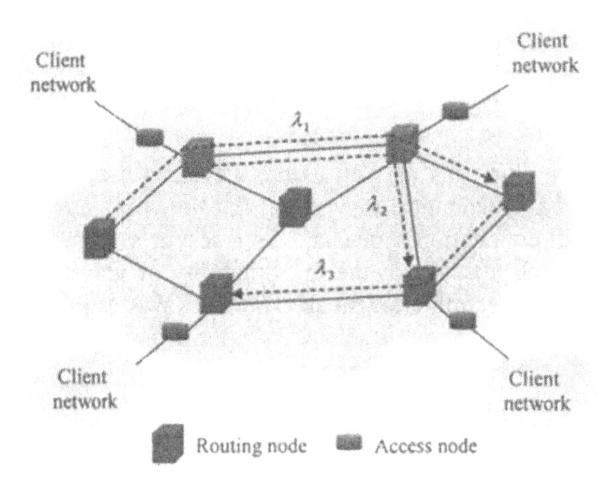

Figure 3.1 A wavelength-routed WDM network.

Example 3.1: Consider a network with three nodes and two wavelengths available on each fiber link, as shown in Figure 3.2. There is no wavelength converter at each node. Suppose that there is one connection already established between node *a* and node *b* on wavelength 1 (λ_1) and between node *b* and node *c* on wavelength 2 (λ_2), respectively. Now there is a connection request between node *a* and node *c*. In this case, there is a wavelength available on each of the fiber links between node *a* and node *c*, i.e., λ_2 between node *a* and node *b* and λ_1 between node *b* and node *c*. However, because the two wavelengths are not continuous, the connection request cannot be accommodated and will accordingly be blocked because of the wavelength-continuity constraint. However, if a wavelength converter is employed at node *b*, the connection request can be accommodated and a lightpath will be established between node *a* and node *c*, using λ_2 between node *a* and node *b* and λ_1 between node *b* and node *c*.

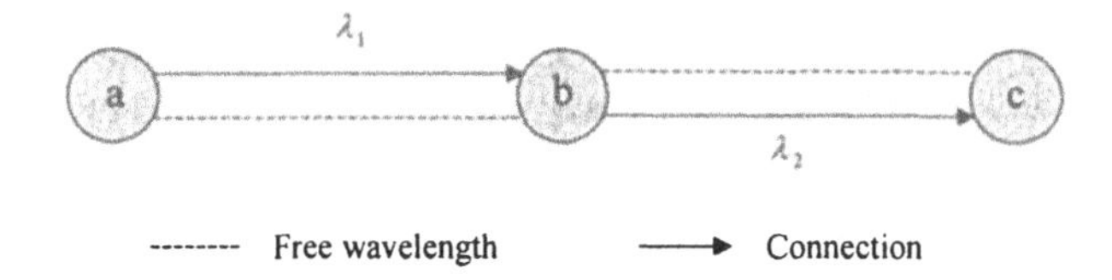

Figure 3.2 Effect of the wavelength-continuity constraint.

A wavelength-routed WDM network is a circuit-switched network in which a lightpath must first be established between a pair of nodes for each connection request before data traffic can actually be transferred. A lightpath is uniquely identified by a physical route and a particular wavelength. To establish a lightpath for a connection request, the network must first decide a physical route from the source node to the destination node and then assign an available wavelength on each link of the decided route. The problem of routing and assigning a wavelength to establish a lightpath is referred to as the routing and wavelength assignment (RWA) problem, and an algorithm that is used to implement RWA is referred to as an RWA algorithm.

3.2.1 Static RWA and Dynamic RWA

Typically, there are two types of network traffic: static traffic and dynamic traffic. For static traffic, a set of connection requests is known a priori. The objective of the RWA problem is to establish a set of lightpaths to accommodate all the connection requests and meanwhile minimize the number of wavelengths used in the network. In other words, the objective is to establish as many lightpaths as possible for a given number of wavelengths on each fiber link. The static RWA problem is also known as the static lightpath establishment (SLE) problem. This problem can be formulated as an integer linear programming (ILP) problem [1], which has proved to be NP-complete [2], and is therefore computationally intractable. It usually takes a large amount of time to find a solution to the problem. However, once a set of connection requests are known, it will not change in the short run. A relatively sufficient amount of time is allowed to obtain an optimal or approximate optimal solution. Therefore, a static RWA algorithm can be performed off-line. The SLE problem will be discussed in more detail in Section 3.3.

For dynamic traffic, connection requests arrive to and depart from the network dynamically in a random manner. A lightpath is established when there is a connection request and is released when the data transfer is completed. The objective of the RWA problem is to route and assign wavelengths in a manner that can minimize the request blocking probability in the network or maximize the network throughput in terms of the number of lightpaths established in the network. The dynamic RWA problem is also referred to as the dynamic lightpath establishment (DLE) problem. Unlike for the SLE problem, a dynamic RWA algorithm must establish a lightpath on the arrival of a connection request and therefore must be performed on-line. For this reason, it must be computationally as simple as possible. The DLE problem will be discussed in more detail in Section 3.4.

For both static and dynamic RWA problems, the combined RWA problem can be computationally hard to handle in many situations, especially for networks of large sizes. An effective way to make the problem more tractable is to divide the combined problem into a couple of subproblems, the routing subproblem and the wavelength assignment subproblem, and then solve the two subproblems separately. For both subproblems, a variety of algorithms have been proposed in the literature [3].

3.2.2 Centralized RWA and Distributed RWA

RWA can be performed under either centralized control or distributed control. Under centralized control, there is a centralized entity or centralized controller in the network. The centralized controller maintains global network state information and is responsible for lightpath establishment on behalf of all network nodes. For static traffic, the centralized controller performs a RWA algorithm to establish a set of lightpaths for a given set of connection requests. For dynamic traffic, all connection requests arrive to the network randomly and are sent to the centralized controller to be processed one by one in a sequential manner. Obviously, centralized control is relatively simple to implement and works well for static traffic. However, it is not scalable and reliable and may become the bottleneck of the network. A failure with the central controller may result in the breakdown of the entire network. As a result, centralized control is considered unsuitable for large networks with dynamic traffic.

Under distributed control, all connection requests are processed at different network nodes concurrently and each node makes its decisions independently based on the network state information it maintains. Compared with centralized control, distributed control improves the scalability and reliability of the network, and is therefore highly preferred for large networks with dynamic traffic. However, a major problem that arises with distributed control is the increased difficulty and complexity in control and management. This chapter focuses on centralized RWA. Distributed RWA will be discussed in Chapter 5.

3.3 Static RWA

In this section, we discuss the static RWA problem, which is also referred to as the SLE problem. Given the physical topology of a network and a set of connection requests, the problem of routing and assigning wavelengths to establish a set of lightpaths over the physical topology to accommodate the set of connection requests is referred to as the static RWA problem or SLE problem. For the SLE problem, a set of connection requests is known a priori. The objective of the SLE problem is to minimize the number of wavelengths needed to establish lightpaths for a given set of connection requests or to maximize the number of lightpaths established for a given set of connection requests and a given number of wavelengths. In general, the SLE problem can be considered as an optimization problem and can be formulated as an ILP problem. In this section, we will present several typical ILP formulations that have already been proposed in the literature for the static RWA problem.

3.3.1 ILP Formulations for SLE Without Wavelength Conversion

In this section, we present two ILP formulations for SLE without wavelength conversion.

Formulation 3.1

The SLE problem is to minimize the number of wavelengths needed to establish lightpaths for a given set of connection requests, which corresponds to minimizing the number of lightpaths established on each link. This problem can be formulated as an ILP problem, which is described as follows. In this formulation, multiple lightpaths are allowed to be established between a pair of source and destination nodes. The notations used are defined as follows.

- λ_{sd}: the number of connection requests from source node s to destination node d
- $c_{sd}(w)$: the number of connections established from source node s to destination node d on wavelength w. Because a wavelength can only be occupied by one connection, $c_{sd}(w)$ is defined as

$$c_{sd}(w) = \begin{cases} 1 & \text{if } w \text{ is occupied} \\ 0 & \text{otherwise} \end{cases}$$

- $c_{ij}^{sd}(w)$: the number of connections established on wavelength w from source node s to destination node d on a link from node i to node j. Because a wavelength on each link can only be assigned to one connection, $c_{ij}^{sd}(w)$ is defined as

$$c_{ij}^{sd}(w) = \begin{cases} 1 & \text{if } w \text{ is assigned} \\ 0 & \text{otherwise} \end{cases}$$

- $C_{\max}^{ij}$: the number of lightpaths established on each link

The optimization objective is to minimize the number of connections established on each link, that is,

Objective:

$$Min[C_{\max}^{ij}] \tag{3-1}$$

subject to

$$C_{\max}^{ij} \geq \sum_{sd} \sum_{w} c_{ij}^{sd}(w), \quad \forall i, \forall j \tag{3-2}$$

$$\sum_i c_{ik}^{sd}(w) - \sum_j c_{kj}^{sd}(w) = \begin{cases} -c_{sd}(w) & if\ \ s = k \\ c_{sd}(w) & if\ \ d = k \\ 0 & otherwise \end{cases} \tag{3-3}$$

$$\sum_w c_{sd}(w) = \lambda_{sd} \tag{3-4}$$

$$\sum_{sd} c_{ij}^{sd}(w) \le 1 \tag{3-5}$$

where equation (3-3) denotes that the connections established on wavelength w from node s to node d are conserved at each node; equation (3-4) specifies that the sum of all connections established from node s to node d is equal to the traffic demand from node s and node d; equation (3-5) ensures that the number of connections established on the link from node i to node j and wavelength w cannot exceed 1.

Formulation 3.2

The dual problem is to maximize the number of lightpaths established in the network for a given number of wavelengths and a given set of connection requests. This problem can also be formulated as an ILP problem [1]. The notations used in this formulation are defined as follows.

- N: the number of source-destination pairs in the network
- M: the number of links in the network
- W: the number of wavelengths available on each link
- P: the number of paths that a connection can be routed through
- ρ: the total number of connection requests or the offered load
- m_i: the number of connections established for source-destination pair i (i=1, 2, ..., N)
- p_i: the fraction of the load for source-destination pair i (i=1, 2, ..., N)
- **m**: a $1 \times N$ vector defined as follows:

$$\mathbf{m} = (m_i,\ i = 1,\ 2,\ \ldots,\ N)$$

- **p**: a $1 \times N$ vector defined as follows:

$$\mathbf{p} = (p_i,\ i = 1,\ 2,\ \ldots,\ N)$$

- **A**=(a_{ij}): a $P \times N$ matrix, in which a_{ij} is defined as

$$a_{ij} = \begin{cases} 1 & if\ \ path\ i\ is\ between\ source-destination\ pair\ j \\ 0 & otherwise \end{cases}$$

- **B**=(b_{ij}): a $P \times M$ matrix, in which b_{ij} is defined as

$$b_{ij} = \begin{cases} 1 & \textit{if link } j \textit{ is on path } i \\ 0 & \textit{otherwise} \end{cases}$$

- **C**=(c_{ij}): a $P \times M$ matrix, in which c_{ij} is defined as

$$c_{ij} = \begin{cases} 1 & \textit{if wavelength } j \textit{ is assigned to path } i \\ 0 & \textit{otherwise} \end{cases}$$

- C(ρ, **p**): the number of connections established in the network

The optimization objective is to maximize the number of connections established in the network, that is,

Objective:

$$Max[C(\rho, \mathbf{p}) = \sum_{i=1}^{N} m_i] \tag{3-6}$$

subject to

$$m_i \geq 0, \quad \text{integer}, \quad i = 1, 2, \ldots, N \tag{3-7}$$

$$c_{ij} \geq 0, \quad \text{integer}, \quad i = 1, 2, \ldots, P \quad j = 1, 2, \ldots, W \tag{3-8}$$

$$\mathbf{C}^T \mathbf{B} \leq \mathbf{1}_{W \times M} \tag{3-9}$$

$$\mathbf{m} \leq \mathbf{1}_W \mathbf{C}^T \mathbf{A} \tag{3-10}$$

$$m_i \leq p_i \rho \quad i = 1, 2, \ldots, N \tag{3-11}$$

where $\mathbf{1}_{X \times Y}$ denotes an $X \times Y$ matrix in which all elements are unity and $\mathbf{1}_X$ denotes a $1 \times X$ matrix in which all elements are unity. Equation (3-9) specifies that a wavelength can be used at most once on each fiber link; equations (3-10) and (3-11) ensure that the number of connections established is less than the number of connections requested.

3.3.2 ILP Formulation for SLE with Wavelength Conversion

As already mentioned, the wavelength-continuity constraint would largely reduce wavelength utilization and therefore degrade network performance. To eliminate the wavelength-continuity constraint, wavelength converters must be deployed at the routing nodes to convert one wavelength to another wavelength. If a wavelength converter is capable of converting any wavelength to any other wavelength, it is said to have a full degree of wavelength conversion. A network with a full degree of wavelength conversion at each node is equivalent to a conventional circuit-switched network. For such networks, only the routing problem needs to be addressed and wavelength assignment is no longer a problem.

Formulation 3.3

Similar to Formulation 3.1, the SLE problem with wavelength conversion can be formulated as follows [3]. Let c_{ij}^{sd} be the number of connections established from source node s to destination node d on a link from node i to node j, and $C_{\max}^{ij}$ be the number of connections established on each link. The optimization objective is to minimize the number of connections established on each link, that is,

Objective:

$$Min[C_{\max}^{ij}] \tag{3-12}$$

subject to:

$$C_{\max}^{ij} \geq \sum_{sd} c_{ij}^{sd} \tag{3-13}$$

$$\sum_{i} c_{ik}^{sd} - \sum_{j} c_{kj}^{sd} = \begin{cases} -\lambda_{sd} & \text{if } s = k \\ \lambda_{sd} & \text{if } d = k \\ 0 & \text{otherwise} \end{cases} \tag{3-14}$$

where equation (3-14) denotes that the connections established between source node s and destination node d are conserved at each node.

Formulation 3.4

Similar to Formulation 3.2, the dual problem with wavelength conversion can be formulated as follows. Let $\mathbf{c}=(c_i)$ be a vector, where c_i denotes the number of connections established on path i, and $C'(\rho, \mathbf{p})$ be the number of connections established in the network. The optimization objective is to maximize the number of connections established in the network, that is,

Objective:

$$Max[C'(\rho, \mathbf{p}) = \sum_{i=1}^{N} m_i] \tag{3-15}$$

subject to

$$m_i \geq 0, \quad \text{integer}, \qquad i = 1, 2, \ldots, N \tag{3-16}$$

$$c_i \geq 0, \quad \text{integer}, \qquad i = 1, 2, \ldots, P \tag{3-17}$$

$$\mathbf{cB} \leq \mathbf{1}_M W \tag{3-18}$$

$$\mathbf{m} \leq \mathbf{cA} \tag{3-19}$$

$$m_i \leq p_i \rho \qquad i = 1, 2, \ldots, N \tag{3-20}$$

where equation (3-18) specifies that the number of wavelengths used on each fiber link cannot exceed W; equation (3-19) specifies that the number of connections established for a node pair does not exceed the sum of connections established on each path of the node pair; equation (3-20) ensures that the number of connections established for a node pair does not exceed the number of requests for the node pair.

3.3.3 ILP Formulation for Static Routing

The combined RWA problem has proved to be an NP-hard problem and thus is computationally difficult to handle. To make the problem more tractable, we can approximately divide the combined problem into two independent subproblems: the routing subproblem and the wavelength assignment subproblem [4]. The routing subproblem can also be formulated as an ILP problem, in which the optimization objective is to minimize the number of connections established on each fiber link. This formulation is different from Formulation 3.1 in that it does not impose the wavelength-continuity constraint. As a result, the ILP formulation for static routing is the same as Formulation 3.3.

3.3.4 Static Wavelength Assignment

Once the routing subproblem is solved, a route has been decided for each connection. Accordingly, the remaining problem is to consider the wavelength assignment subproblem, i.e., to assign an available wavelength to each connection such that two lightpaths are not assigned the same wavelength on a given link and in the absence of wavelength conversion, a lightpath is assigned the same wavelength on all the links it traverses. It is interesting that the wavelength assignment problem turns out to be a problem that is closely related to the graph-coloring problem [1–2].

To explain the graph-color problem, consider a graph *G*, which represents a network. Each vertex in graph *G* represents a node in the network, and each undirected edge between two vertices corresponds to a fiber link between two nodes. Further, the route for a lightpath corresponds to a path in graph G, and thus a set of routes in the network corresponds to a set of paths in graph *G*. Now construct an auxiliary graph *A(G)*, such that each node in graph *A(G)* corresponds to a path in graph *G*. If two paths share a common edge in graph *G*, the corresponding nodes in graph *A(G)* are connected by an undirected edge. As a result, the wavelength assignment problem is equivalent to solving the classic graph coloring problem in graph *A(G)*, that is, to assign a color to each node in graph *A(G)* such that no two adjacent nodes have the same color and meanwhile the total number of colors is minimized. These colors correspond to the wavelengths used on the paths in graph G. The minimum number of colors required to color the nodes of a graph in this manner is called the chromatic number of the graph. Therefore, the minimum number of wavelengths required to solve the wavelength assignment problem is the chromatic number of graph *A(G)*.

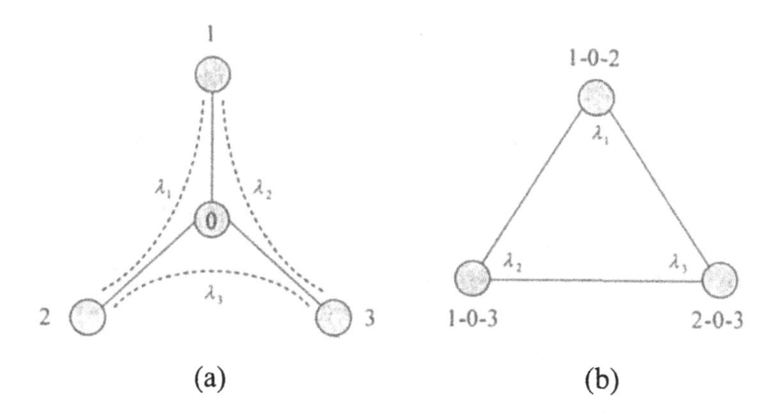

Figure 3.3 Relationship to graph coloring.

Example 3.2: Consider a graph G of a network, as shown in Figure 3.3(a). Figure 3.3(b) depicts its auxiliary graph $A(G)$. Suppose that we only need to establish a lightpath between node *1* and node *2*, node *2* and node *3*, and node *1* and node *3*, respectively. Obviously, the chromatic number of $A(G)$ is three and the minimum number of wavelengths required to solve the wavelength assignment problem is also three.

The graph-coloring problem is a hard problem and is known to be an NP-complete problem [5]. However, there are efficient graph-coloring algorithms for some special types of graphs, which can find an exact solution to the wavelength assignment problem. For a regular graph, unless it has only a few nodes, an exact solution is difficult to find. Only approximate solutions can be found. To address the general graph-coloring problem, a variety of approximate algorithms have been designed [6–7], and these algorithms can be used to obtain approximate solutions to the wavelength assignment problem.

3.4 Dynamic RWA

In this section, we discuss the dynamic RWA problem, which is also referred to as the DLE problem. For dynamic traffic, connection requests arrive to the network dynamically in a random manner, which are sent to a network controller in the network for processing and are processed sequentially one by one. To accommodate the connection requests, the network controller performs a dynamic RWA algorithm to establish a lightpath for each connection request based on the network state information (e.g., network topology and wavelength usage) it maintains. If a lightpath cannot be established for a connection request because of the unavailability of wavelengths, the request will be blocked. To establish a lightpath for each connection request, the network must first decide a physical route from the source node to the destination node and then assign an available wavelength on each link of the decided route, which involves both routing and wavelength assignment. In this section, we first introduce basic routing algorithms and wavelength assignment algorithms and then present several well-known RWA algorithms that have already been proposed in the literature to address the DLE problem. Similar to the static RWA problem, the combined dynamic RWA problem can also be divided into the routing subproblem and the wavelength assignment subproblem.

3.4.1 Routing

There are typically three types of routing paradigms for dynamic routing [3]:

- Fixed routing
- Fixed-alternate routing
- Adaptive routing

Fixed Routing

In fixed routing, there is only a single fixed route for each pair of network nodes. This fixed route is precomputed off-line, and any connection between a pair of source and destination nodes uses the same fixed route, which imposes a strict restriction on route selection. A typical example of fixed routing is fixed shortest-path routing. Figure 3.4 illustrates the fixed shortest-path route from node *a* to node *d*. Obviously, fixed routing is simple to implement and thus has a lower computational complexity. However, because it imposes a strict restriction on routing, it can result in a higher request blocking probability in the network. If there is no wavelength available on any link of the shortest-path route or if there is no common wavelength available on all links of the shortest-path route, a connection request from node *a* to node *d* will be blocked. Moreover, fixed routing has no fault-tolerant capability. If there is a link failure on the shortest-path route, a connection from node *a* to node *d* will be disrupted and blocked.

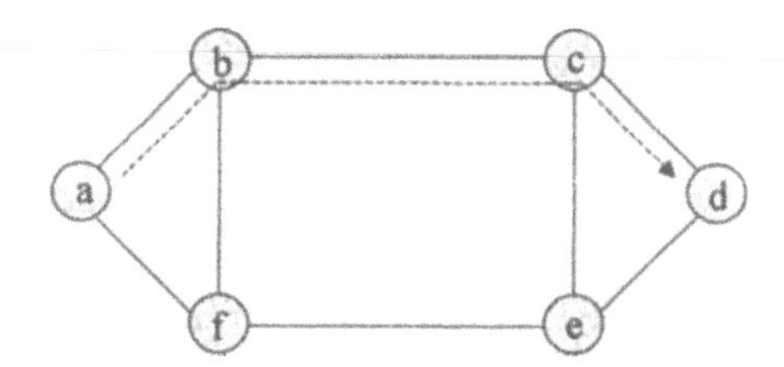

Figure 3.4 Fixed shortest-path routing.

Fixed-Alternate Routing

In fixed-alternate routing, there is a set of alternate routes for each pair of network nodes. The actual route for a connection request can only be chosen from these alternate routes, which also imposes a restriction on route selection. These alternate routes are precomputed off-line and are orderly stored in a routing table maintained by the network controller. For example, they may include the shortest-path route, the second shortest-path route, the third shortest-path route, etc. Usually, they are link-disjoint routes in the sense that any of the routes does not share any link with any other of the routes. In most cases, these routes are ordered in the routing table in terms of the number of link hops between a pair of nodes. The first route in the routing table is the shortest-path route. To accommodate a connection request, the network controller searches the alternate routes in the routing table in sequence until an available route is found. If no available route is found, the connection request will be blocked. Compared with fixed routing, fixed-alternate routing can significantly reduce the request blocking probability without largely increasing the computational complexity in routing. Moreover, it can also provide some degree of fault-tolerant capability. In the event of a link failure, an alternate route can be used to recover the disrupted service. Figure 3.5 illustrates a couple of alternate routes from node *a* to node *d*.

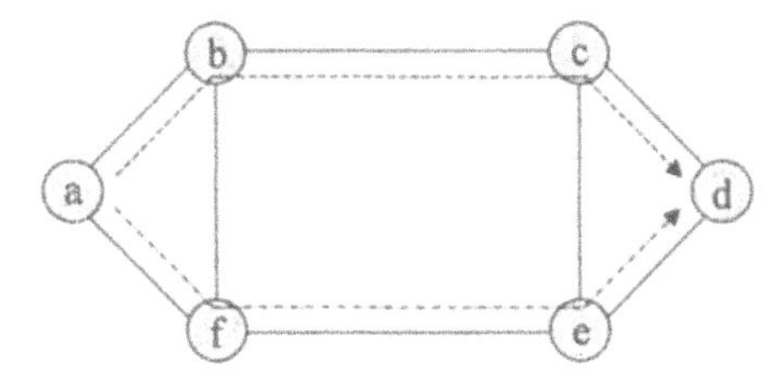

Figure 3.5 Fixed-alternate routing.

Adaptive Routing

In adaptive routing, there is no restriction on route selection. Any possible route between a pair of source and destination nodes can be chosen as an actual route for a connection. The choice of a route is based on the current network state information as well as a path-selection policy, such as the least-cost path first or the least-congested path first. The network controller maintains global network state information and dynamically makes a routing decision for each connection request. Compared with fixed-alternate routing, adaptive routing can further reduce the request blocking probability and provide a higher degree of fault-tolerant capability. However, it may largely increase the computational complexity in making a routing decision. Figure 3.6 illustrates a least-cost route from node *a* to node *d*, where the label on each link presents the cost for using that link.

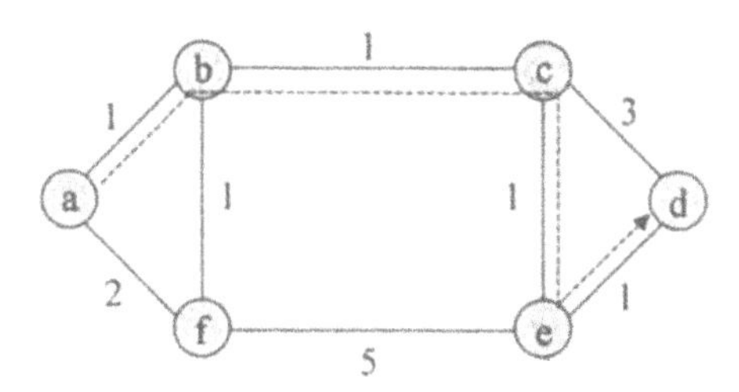

Figure 3.6 Adaptive routing with least-cost path first.

3.4.2 Wavelength Assignment

There are a variety of wavelength assignment algorithms proposed for wavelength assignment, but the most commonly used wavelength assignment algorithms include fixed-ordered, random-ordered, least-used, and most-used [3].

Random (RANDOM)

The Random algorithm searches the wavelengths in a random order. It first determines the wavelengths that are available on each link of a decided route and then chooses an available wavelength randomly among the available wavelengths determined, usually with uniform probability. This algorithm has no communication overhead.

First-Fit (FF)

The First-Fit algorithm searches the wavelengths in a fixed order. All wavelengths are indexed and are searched in the order of their index numbers. The first available wavelength found is chosen. This algorithm has a lower computational cost than the random-ordered algorithm because it does not necessarily determine the wavelengths that are available on each link of a decided route. It also has no communication overhead.

Least-Used (LU)/SPREAD

The Least-Used algorithm chooses the wavelength that is the least used in the network. The purpose is to balance the traffic load over all the wavelengths. The idea behind this algorithm is that a shorter route is more likely to be found on the least-used wavelength than on the most-used wavelength, which can result in more links being available for those connection requests that arrive later. This algorithm requires global network state information to compute the least-used wavelength, which introduces additional communication overhead. Moreover, it also requires additional storage and computation cost. As a result, this algorithm is more suitable for centralized control but not preferred in practice for distributed control.

Most-Used (MU)/PACK

The Most-Used algorithm is just the opposite of the Least-Used algorithm. It chooses the wavelength that is the most used in the network, thus packing connections into fewer wavelengths. This algorithm also requires global network state information to compute the most-used wavelength. It has communication overhead and storage and computational cost similar to those with the Least-Used algorithm. However, it outperforms the Least-Used algorithm significantly [8]. This algorithm is also more suitable for centralized control but not preferred in practice for distributed control.

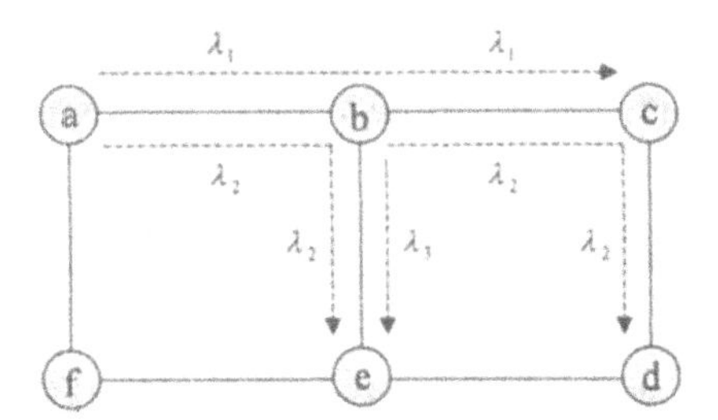

Figure 3.7 Illustration of wavelength assignment.

Example 3.3: Consider a network that has six nodes and seven links with four wavelengths (i.e., λ_1, λ_2, λ_3, and λ_4) on each fiber link, as shown in Figure 3.7. Suppose that there are four connections already established in the network, and wavelengths λ_1, λ_2, λ_3, and λ_4 are used on two, four, one, and zero links, respectively. Now there is a new connection request for a connection from node *a* to node *d*, and *a-f-e-d* is chosen as the route of the connection. In this case, all three wavelengths are available on the chosen route. Accordingly, each of the wavelength assignment algorithms described above will search the wavelengths in a different order and may select a different wavelength for the connection. Specifically, the RANDOM algorithm will search in a random order, say, λ_3, λ_2, λ_1, and λ_4, and will choose wavelength λ_3. The First-Fit algorithm will search in the order λ_1, λ_2, λ_3, and λ_4, and will choose wavelength λ_1 because it is the first available wavelength found. The Least-Used algorithm will search in the order λ_4, λ_3, λ_1, and λ_2, and will choose wavelength λ_4 because it is the least-used wavelength. The Most-Used algorithm will search in the order λ_2, λ_1, λ_3, and λ_4, and will choose wavelength λ_2 because it is the most-used wavelength.

Min-Product (MP)

The Min-Product algorithm [9] was proposed for multifiber networks. The objective is to minimize the number of fibers used in the network. It first computes the value of

$$\prod_{i \in \pi(p)} D_{ij}$$

for each wavelength, where D_{ij} is the number of fibers already used on link i and wavelength j, and $\pi(p)$ is a set of links that constitute path p. Then it chooses the wavelength with the smallest computed value. If there are several wavelengths with the same computed value, the wavelength with the lowest index is chosen. This algorithm introduces additional computational cost. In a single-fiber network, it becomes the First-Fit algorithm.

Example 3.4: Consider a path with six nodes and five links, which has been selected as the route for a connection. Each link has 10 fibers with 3 wavelengths on each fiber. Suppose that D_{ij} (i=0, 1, ..., 4; j=1, 2, 3) is already known, as shown in Table 3.1. Hence, the product value of D_{ij} for each wavelength can be computed and wavelength 3 is selected because it has the smallest computed value.

Table 3.1

j \ i	D_{ij}					$\prod_{i \in \pi(p)} D_{ij}$
	0	*1*	*2*	*3*	*4*	
1	2	3	5	1	6	180
2	3	2	2	5	4	240
3	1	3	2	1	2	12

Least-Loaded (LL)

The Least-Loaded algorithm [10] was also proposed for multifiber networks. It chooses the wavelength that has the largest residual capacity on the most-loaded link along a path. The residual capacity of wavelength j on link i is defined as the number of fibers on which wavelength j is still available. Thus the algorithm chooses the lowest-indexed wavelength j that satisfies

$$\max_{j \in \omega(p)} \{ \min_{i \in \pi(p)} (M_i - D_{ij}) \}$$

where M_i is the number of fibers on link i and $\omega(p)$ is a set of available wavelengths along path p. If used in single-fiber networks, the residual capacity is either 1 or 0. The lowest-indexed wavelength with residual capacity 1 is chosen. Accordingly, this algorithm reduces to the First-Fit algorithm in single-fiber networks. It has been shown in [10] that the Least-Loaded algorithm outperforms the Most-Used algorithm and the First-Fit algorithm in terms of the blocking probability in a multifiber network.

> ***Example 3.5***: Consider a path that consists of six nodes and five links with ten fibers on each link and three wavelengths (i.e., λ_1, λ_2, and λ_3) on each fiber. Suppose that D_{ij} (i=0, 1, ..., 4; j=1, 2, 3) is already known and the value of $(M_{ij}-D_{ij})$ is accordingly known, as shown in Table 3.2. In this case, the most-loaded link for wavelength 1 is link 4, on which the residual capacity of wavelength 1 is 4. The most-loaded link for wavelength 2 is link 3, on which the residual capacity of wavelength 3 is 5. The most-loaded link for wavelength 3 is link 1, on which the residual capacity of wavelength 1 is 7. Accordingly, wavelength 1 is chosen as it has the largest residual capacity on the most-loaded link.

Table 3.2

j \ i	D_{ij}					$M_i - D_{ij}$					$\min_{i \in \pi(p)}$
	0	*1*	*2*	*3*	*4*	*0*	*1*	*2*	*3*	*4*	$(M_i - D_{ij})$
1	2	3	5	1	6	8	7	5	9	4	4
2	3	2	2	5	4	7	8	8	5	6	5
3	1	3	2	1	2	9	7	8	9	8	7

MAX-SUM

The MAX-SUM algorithm [11] was proposed for a fixed-routing network with no wavelength conversion and considers the wavelength assignment problem for dynamic traffic. It assumes that the network is in an arbitrary state φ, in which a set of lightpaths have already been established and wavelengths have already been assigned to these lightpaths. Suppose that there is a new connection request for a lightpath and the route for the lightpath is path p. The objective is to assign an available wavelength to the lightpath so that network performance in terms of the blocking probability can be improved. For this purpose, the algorithm attempts to maximize the total residual path capacity in the network after the lightpath is established.

The path capacity is defined in terms of link capacities. The link capacity of link i on wavelength j, $C(\varphi, i, j)$, is defined as the number of fibers on which wavelength j on link i is still available, i.e.,

$$C(\varphi,i,j) = M_i - D_{ij}(\varphi)$$

$D_{ij}(\varphi)$ is the D_{ij} value in state φ. The capacity of path p on wavelength j in state φ, $C(\varphi, p, j)$, is then defined as the number of fibers on which wavelength j is available on the most-congested link along the path, i.e.,

$$C(\varphi,p,j) = \min_{i\in\pi(p)} C(\varphi,i,j)$$

The path capacity of path p in state φ, $C(\varphi, p)$, is defined as the sum of the path capacities on all wavelengths on a fiber, i.e.,

$$C(\varphi,p) = \sum_{j=1}^{W} \min_{i\in\pi(p)} C(\varphi,i,j)$$

where W is the number of wavelengths on each fiber. Let $\omega(\varphi, p)$ be a set of possible wavelengths that are available for the connection request and let $\varphi'(j)$ be the next network state if wavelength j is assigned to the lightpath. The MAX-SUM algorithm chooses the wavelength j from $\omega(\varphi, p)$ such that it maximizes the total residual path capacity, i.e.,

$$\sum_{p\in P} C[\varphi'(j),p]$$

where P is a set of all possible paths in the network.

Relative-Capacity-Loss (RCL)

The Relative-Capacity-Loss algorithm [12] is based on the MAX-SUM algorithm. It is not difficult to see that maximizing the total residual path capacity is equivalent to minimizing the total capacity loss on all possible lightpaths, which is defined as

$$\sum_{p\in P} \{C(\varphi,p) - C[\varphi'(j),p]\}$$

Because only the path capacity on wavelength j will change after a lightpath is established on wavelength j, the MAX-SUM algorithm actually chooses wavelength j to minimize the total capacity loss on wavelength j, i.e.,

$$\sum_{p \in P} \{C(\varphi, p, j) - C[\varphi'(j), p, j]\}$$

However, minimizing the total capacity loss sometimes does not lead to the best performance. For this reason, the Relative-Capacity-Loss algorithm attempts to choose wavelength j to minimize the relative capacity loss, which is defined as

$$\sum_{p \in P} \frac{C(\varphi, p, j) - C[\varphi'(j), p, j]}{C(\varphi, p, j)}$$

It has been shown in [12] that in most cases the Relative-Capacity-Loss algorithm performs better than the MAX-SUM algorithm.

The wavelength assignment algorithms described thus far aim at minimizing the overall blocking probability in the network. However, this may result in a connection request for a longer connection to be more likely blocked. For this reason, some wavelength assignment strategies have been proposed to protect longer connections, such as Wavelength-Reservation and Threshold-Protection [13].

Wavelength-Reservation (Rev)

With the Wavelength-Reservation strategy, a given wavelength is reserved on a specific link for a longer connection or multihop connection. A shorter connection cannot get this wavelength even though the wavelength is free. As a result, the blocking probability for longer connections could be reduced while that for shorter connections would be increased.

Threshold-Protection (Thr)

With the Threshold-Protection strategy, a single-hop or shorter connection is assigned a wavelength on a specific link only if the number of free wavelengths on the link is at or above a given threshold. If the number of wavelengths is below the threshold, no wavelength can be assigned to a shorter connection. Obviously, this would also increase the blocking probability for shorter connections but could reduce the blocking probability for longer connections.

The Wavelength-Reservation and Threshold-Protection strategies do not specify which wavelength to choose for a given connection request. Instead, they just specify for what connections or under what conditions a wavelength can be assigned to a given connection request. For this reason, they must be used in combination with a wavelength assignment algorithm. On the other hand, unlike the wavelength assignment algorithms that aim at minimizing the overall blocking probability in the network, both strategies only attempt to reduce the blocking probability for longer or multihop connections, and thus may result in a higher overall blocking probability in the network. In Section 3.5, we will discuss both strategies again from the viewpoint of fairness.

3.4.3 RWA algorithms

We now present several well-known RWA algorithms that have already been proposed for dynamic traffic, including the fixed shortest-path routing algorithm, the *K*-shortest-path routing algorithm, the least-congested-path routing algorithm, and the least-cost-path routing algorithm.

Fixed shortest-path routing algorithm

The fixed shortest-path routing algorithm uses the shortest-path route for any connection between a pair of source and destination nodes. The shortest-path route is precomputed off-line by using a shortest-path algorithm, such as *Dijkstra*'s algorithm [14] or the Bellman-Ford algorithm [15]. When a connection request arrives to the network, this algorithm examines the wavelength availability on each link of the shortest-path route from the source node to the destination node of the connection. If there is one common wavelength available on all the links, the connection will be established on the wavelength. If no wavelength is available on any of the links or there is no common wavelength on all the links, the connection request will be blocked. If multiple common wavelengths are available, a connection is established on one of them, which is selected with a wavelength assignment algorithm, such as Random or First-Fit.

The main advantage of this algorithm is its simplicity in implementation and shorter connection setup time because it always chooses the shortest-path route. However, it may result in a higher blocking probability because of the strict restriction on routing. For a pair of network nodes, there are many cases in which no wavelength is available on the shortest-path route but some wavelengths are available on other routes. In such cases, a connection request will still be blocked. For this reason, this algorithm can only perform better under light traffic load. As the traffic load increases, the performance of this algorithm will largely degrade. On the other hand, the network topology also has a great impact on the performance of this algorithm. If the network is densely connected, the performance of this algorithm will largely degrade.

K-shortest-path routing algorithm

The K-shortest-path routing algorithm [16–17] selects the shortest-path route among the first K shortest-path routes for each pair of source and destination nodes. These routes are stored in the routing table and are examined in the order of their distances from the shortest-path route to the longest-path route. The distance can be measured in terms of the number of link hops or the end-to-end propagation delay. When a connection request arrives to the network, this algorithm examines the first shortest-path route first. If no wavelength is available on this route, the next shortest-path route is examined. The first available route is chosen as the actual route. Note that in the absence of wavelength conversion, a route is available in the sense that there is at least one common wavelength available on all the links of the route. If multiple wavelengths are available on the route, a wavelength assignment algorithm is used to select one. If no route is available, the request will be blocked.

The main advantage of this algorithm is its simplicity in implementation, shorter connection setup time, and reduced request blocking probability compared with the fixed shortest-path routing algorithm. However, because it also imposes some degree of restriction on routing, it cannot achieve the best network performance in terms of the blocking probability.

Least-congested-path routing algorithm

The least-congested-path routing algorithm selects the route with the least congestion among all the possible routes for a pair of source and destination nodes. The congestion on a link is measured in terms of the number of wavelengths available on the link. The fewer the number of available wavelengths, the more congested the link. The congestion on a route is measured in terms of the number of wavelengths available on the most congested link of the route. This algorithm can be based on either fixed-alternate routing or adaptive routing. With fixed-alternate routing, a set of K alternate routes is precomputed off-line as the possible routes [18–19]. With adaptive routing, all the possible routes for a pair of nodes can be considered. The main difference between fixed-alternate routing and adaptive routing lies in the number of possible routes. Fixed-alternate routing allows only a limited number of alternate routes, whereas adaptive routing has no restriction on the number of possible routes. When a connection request arrives to the network, the algorithm will first compute the congestion on the possible routes and then select the least-congested route as the actual route. If multiple routes have the same congestion, the shortest-path route will be selected. Once a route is selected, a wavelength assignment algorithm is used to assign an available wavelength for the connection.

> ***Example 3.6***: Consider again the network in Figure 3.7. Suppose that the wavelength usage on each link is shown in Figure 3.8. Now there is a new connection request for a lightpath from node *a* to node *d*. To establish this lightpath, there are four possible routes: *a-b-c-d*, *a-b-e-d*, *a-f-e-d*, and *a-f-e-b-c-d*, and the congestions on these routes are 3, 3, 2, and 3, respectively. As a result, the least-congested-path routing algorithm will select *a-f-e-d* as the route to establish the lightpath.

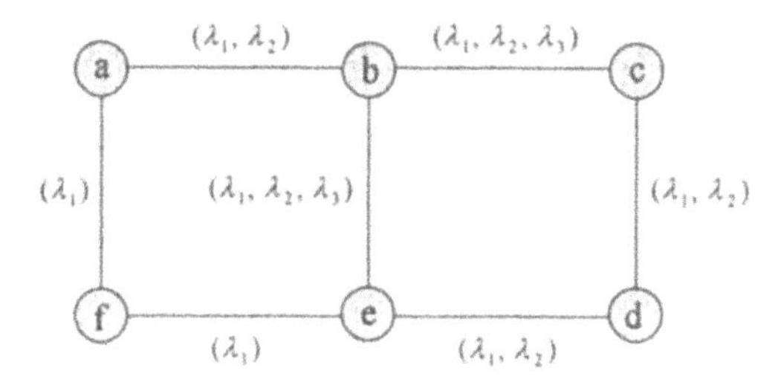

Figure 3.8 Least-congested-path routing.

Least-cost-path routing algorithm

The least-cost-path routing algorithm selects the route with the least cost among all the possible routes for a pair of source and destination nodes. The cost of a route is usually measured in terms of hop count, end-to-end delay, or resource availability. The hop count of a route is the number of links on the route. If the hop count is used to measure the cost of a route, any link has the same cost. If the distance of a route must be reflected in the cost, the propagation delay can be used to measure the cost of the route. The end-to-end delay is equal to the total delay on all links of the route, which is proportional to the distance of the route. In addition, the cost of a route can also be measured in terms of the availability of network resources, such as wavelengths, and the cost of devices used at intermediate nodes, such as wavelength converters. The more resources are available, the lower the cost. The more expensive a device is, the higher the cost of using the device. The *K*-shortest-path routing algorithm is an example using hop count or end-to-end delay to measure the cost. The fixed-alternate least-congested-path routing algorithm is an example using the wavelength availability to measure the cost. This algorithm can also be based on either fixed-alternate routing or adaptive routing. With fixed-alternate routing, a set of K alternate routes are precomputed off-line as the possible routes. With adaptive routing, all the possible routes for a pair of nodes can be considered. When a connection request arrives to the network, the algorithm will first compute the cost of the possible routes and then select the least-cost path as the actual route. If multiple routes have the same cost, the shortest-path route is selected. Once a route is selected, a wavelength assignment algorithm is used to assign an available wavelength for the connection.

> ***Example 3.7***: Consider again the network in Figure 3.7. Suppose that each link with available wavelengths has a cost of 1. Each link without available wavelengths has a cost of ∞. Each link with wavelength converters has a cost of c (>1). The cost on each link is shown in Figure 3.9. Now there is a new connection request for a lightpath from node *a* to node *d*. To establish this lightpath, there are four possible routes: *a-b-c-d*, *a-b-e-d*, *a-f-e-d*, and *a-f-e-b-c-d*, which have a cost of 3, ∞, $2c+1$, and ∞, respectively. As a result, the least-cost-path routing algorithm will select *a-b-c-d* as the route to establish the lightpath.

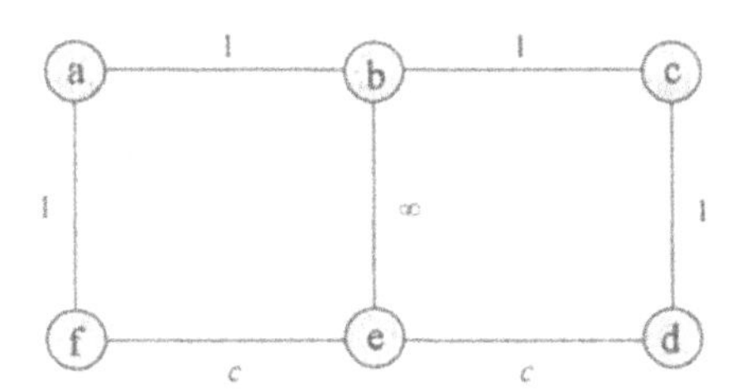

Figure 3.9 Least-cost-path routing.

3.5 RWA for Fairness

Fairness is an important problem that must be considered in routing and wavelength assignment. In a practical network, different network users need to establish connections between network nodes with different distances or hop counts. The network should accommodate different network users or connection requests in a fair manner. Here the hop count of a connection is defined as the number of physical links on the shortest path between the source node and the destination node of the connection. However, because of the wavelength constraints inherent in the network, in particular the wavelength-continuity constraint, a RWA algorithm usually tends to establish a shorter connection with a smaller hop count more easily than a longer connection with a larger hop count. In other words, a longer connection request with a larger hop count is more likely to be blocked than a shorter connection request with a smaller hop count because of insufficient wavelength resources in the network. This may result in a big performance difference in terms of the request blocking probability for different connection requests and is therefore unfair for different network users. To improve the fairness among different network users, the network must take appropriate measures to effectively reduce or minimize the difference in terms of the request blocking probability for connection requests with different hop counts.

One simple method to address the fairness problem is to employ wavelength converters at a network node. Wavelength converters can alleviate and even eliminate the wavelength-continuity constraint, and therefore significantly reduce the request blocking probability for longer connections with larger hop counts. However, this method cannot thoroughly solve the problem, because shorter connections with smaller hop counts can benefit from wavelength converters as well. Moreover, the deployment of wavelength converters would substantially increase the cost and complexity of the network. To improve fairness, another possible method is to perform wavelength rerouting. By rerouting some connections already existing in the network, this method can make a wavelength-continuous route that is initially not wavelength continuous for a new connection request. However, this also cannot solve the fairness problem thoroughly. It would largely increase the complexity in control and may incur significant service disruptions in the network. For these reasons, it is necessary to take the fairness problem into account in RWA and an RWA algorithm that is capable of improving fairness is highly desirable.

It should be pointed out that fairness is usually improved at the cost of overall network performance in terms of request blocking probability. Accordingly, any method for improving fairness must keep degradation of the overall network performance under an acceptable level. In the next sections, we will introduce three different methods that have been proposed in the literature for improving fairness in RWA [13][17], including wavelength reservation, threshold protection, and limited alternate routing.

3.5.1 Wavelength Reservation

In the wavelength reservation method [13], one or more wavelengths are reserved on each link of a longer-hop route, exclusively for a longer connection that uses the route. This longer connection is allowed to use other wavelengths on each link of the route, but other connections are not allowed to use the wavelengths reserved for this longer connection. As a result, the blocking probability for this connection can be significantly reduced. However, this may also reduce the wavelength utilization and increase the overall blocking probability in the network. For example, if one or more wavelengths are reserved on each link of a route but the number of connections that use the route is very small, the reserved wavelengths are not in use most of the time. In this case, the reserved wavelengths cannot be used efficiently, which would significantly reduce the wavelength utilization and therefore increase the overall blocking probability in the network. Therefore, there is a trade-off between fairness and performance in addressing the fairness problem.

3.5.2 Threshold Protection

In the threshold protection method [13], the shorter connections with smaller hop counts are assigned an available wavelength on each link only if the number of available wavelengths on that link is above a given threshold. The purpose is to provide a certain number of available wavelengths on each link for use by longer connections with larger hop counts. Accordingly, this can largely reduce the blocking probability as well as the setup time for the longer connections. However, the problem that arises with this method is that the wavelengths that remain available on each link may not be wavelength continuous. This would, to a large extent, reduce the benefit to the longer connections. In particular, if there is only a small amount of traffic for longer connections, this can only result in unnecessary degradation in the performance of shorter connections and overall network performance.

3.5.3 Limited Alternate Routing

In the limited alternate routing method [17], a shorter connection with a smaller hop count is provided with fewer alternate routes whereas a longer connection with a larger hop count is provided with more alternate routes in proportion to the number of hop counts. The purpose is to provide more choices of routes for longer connections and thus accommodate more connections with larger hop counts. If the network is densely connected and the traffic load is light, this method can effectively improve fairness among connections with different hop counts as well as overall network performance. This is because in this case additional routes provide more chances to find an available wavelength on a route. However, if the network is sparsely connected and the traffic load is high, additional routes do not mean more chances to find an available wavelength on a route. Accordingly, this method is unable to provide better fairness than any other method not considering fairness.

In addition to the above three methods, we can also find some other methods in the literature, such as the static priority method, and the dynamic priority method [20].

3.6 Wavelength Rerouting

In this section, we discuss the wavelength rerouting problem, introduce basic lightpath migration operations and rerouting schemes, and present a well-known wavelength rerouting algorithm proposed in the literature.

3.6.1 Need for Wavelength Rerouting

In a wavelength-routed WDM network with no wavelength conversion, a lightpath must use the same wavelength on all the links it traverses, which is referred to as the wavelength-continuity constraint. With this constraint, a connection request is more likely to be blocked because of the unavailability of a common wavelength on all the links of a decided route. This would largely reduce wavelength utilization and thus degrade network performance in terms of blocking probability, in particular, for dynamic traffic. One way to increase wavelength utilization is to deploy wavelength converters at the network nodes. As we mentioned above, wavelength converters can alleviate and even eliminate the wavelength-continuity constraint, and can therefore significantly increase the wavelength utilization and reduce the blocking probability in the network. However, wavelength converters are still very expensive optical devices. The deployment of wavelength converters would greatly increase the cost of the network and thus is not a cost-effective solution. For this reason, wavelength rerouting becomes an effective alternative. By rerouting some connections already existing in the network, wavelength rerouting can create a wavelength-continuous route that is initially not available for a new connection request, and can thus improve network performance in terms of wavelength utilization and blocking probability. To accommodate a new connection request, the network usually performs an RWA algorithm to decide a route for the request and select an available wavelength that is continuous on each link of the decided route. However, it is not always possible to find a wavelength-continuous route at the time of the connection request. In particular, there may be a case in which a route is available but there is no common wavelength on each link of the route. In this case, the connection request is normally blocked. With wavelength rerouting, however, the network will perform a wavelength-rerouting algorithm to see whether it is possible to create a wavelength-continuous route for the connection request by adjusting the wavelengths of some existing lightpaths in the network. If such a route can be created, the connection request will not be blocked and a lightpath can be established for the connection request on the created route. The following example illustrates the benefit of wavelength rerouting.

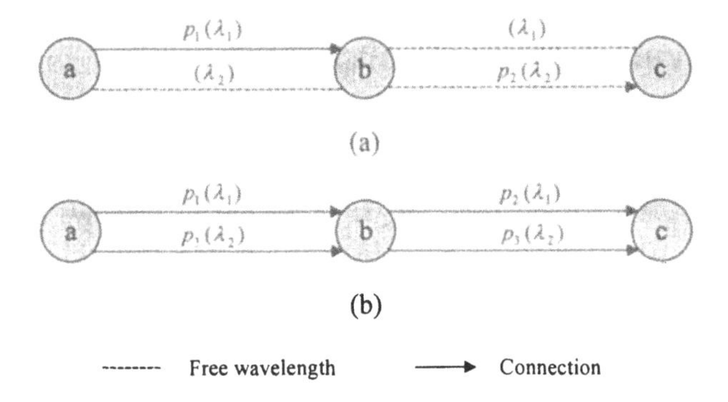

Figure 3.10 Illustration of wavelength-rerouting benefit.

Example 3.8: Consider a route consisting of three nodes and two links with two wavelengths, λ_1 and λ_2, on each link, as shown in Figure 3.10. Suppose that lightpath p_1 has already been established between node *a* and node *b* using wavelength 1 and lightpath p_2 between node *b* and node *c* using wavelength 2. There is a new connection request for a lightpath from node *a* to node *b*. In this case, it is not possible to establish a lightpath along route *a-b-c* for the connection request, although wavelength 1 is still available on link (*b, c*) and wavelength 2 is available on link (*a, b*). However, if we move lightpath p_2 to wavelength 1, wavelength 2 will become available on both link (*a, b*) and link (*b, c*). Then a lightpath p_3 can be established on route *a-b-c* using wavelength 2 for the connection request.

In addition to its use in improving wavelength utilization, wavelength rerouting can be used for handling network failures, such as a node fault and a link cut. In the event of a network failure, all the lightpaths that are affected by the failure will be disrupted. To recover the services on the disrupted lightpaths, a new lightpath must be established for each disrupted lightpath on an end-to-end basis. If there is no wavelength-continuous route available for a new lightpath, wavelength rerouting can be used to create a wavelength-continuous route if such a route exists. Accordingly, wavelength rerouting can significantly improve the network capability in handling network failures.

3.6.2 Problems in Wavelength Rerouting

In wavelength rerouting, the lightpaths to be rerouted must be temporarily disrupted to prevent data from being lost. Because each lightpath carries a large amount of data traffic, a very short period of disruption would result in a significant amount of throughput loss in the network. In particular, in large networks, the disruption period of a rerouted lightpath may be very long because of the propagation delay for transferring control packets. For this reason, it is imperative to minimize the incurred lightpath disruptions, including not only the number of rerouted lightpaths but also the disruption period of each rerouted lightpath. Wavelength rerouting involves two basic problems:

- Lightpath selection
- Lightpath migration

Lightpath selection

Lightpath selection is selection of a set of rerouted lightpaths to create a wavelength-continuous route for a connection request. This is implemented by performing a rerouting algorithm. Because a rerouting algorithm is performed on-line, it must be simple in terms of computational complexity. In addition, the number of lightpaths to be rerouted must be minimized in order to reduce the incurred throughput loss.

Lightpath migration

Lightpath migration is moving each of the selected lightpaths to another wavelength on the same route or another route. To reduce the throughput loss, it is desirable to minimize the incurred disruption period of each rerouted lightpath.

3.6.3 Lightpath Migration Operations

There are two basic operations for lightpath migration: wavelength-retuning (WR) and move-to-vacant (MTV) [21].

Wavelength-Retuning

Wavelength-retuning retunes the wavelength of a lightpath but maintains its route. It has the following advantages:

- It facilitates rerouting control because the route of a lightpath is not changed and the same routing nodes are used.
- It simplifies the computation for rerouting because only the wavelength of a lightpath needs to be considered.

Move-To-Vacant

Move-to-vacant reroutes a lightpath to a vacant route with no other lightpaths. It has the following advantages:

- It does not interrupt other lightpaths because there are no other lightpaths on the new route.
- It preserves the transmission on the old route during the establishment of the new route, and thus reduces the disruption period of the rerouted lightpath.

A basic operation based on WR and MTV is called move-to-vacant wavelength-rerouting (MTV-WR) [21], which combines the advantages of WR and MTV operations.

Move-To-Vacant Wavelength-Retuning

Move-to-vacant wavelength-retuning moves a lightpath to a free wavelength on the same route and thus largely reduces the disruption period.

> ***Example 3.9***: This example illustrates how the MTV-WR operation is implemented. It closely follows that in [21] and is described as follows.
>
> 1) The network controller first sends a control packet to each intermediate routing node (or switch) along the route of the rerouted lightpath, as shown in Figure 3.11(a). This control packet sets the state of the optical switch at each intermediate node to establish a lightpath on an available wavelength from the source node to the destination node, as shown in Figure 3.11(b). Then the source node prepares to switch data transmission from the old wavelength to the new wavelength.

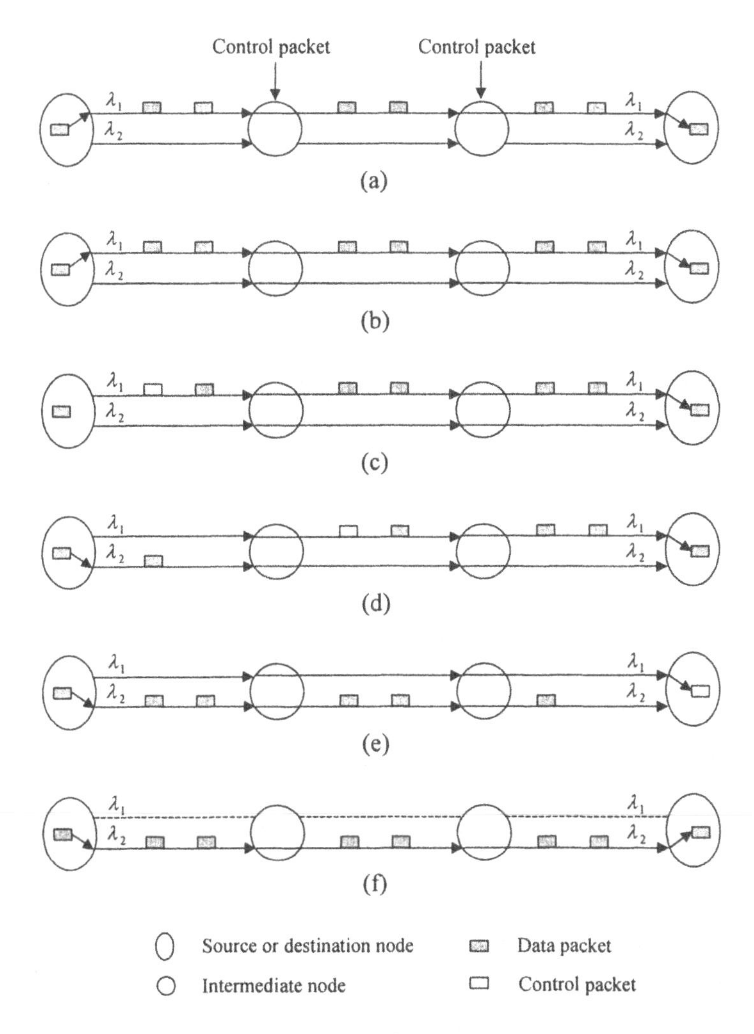

Figure 3.11 Illustration of the MTV-WR operation.

2) The source node then appends an end-of-transmission control packet after the last data packet on the old wavelength and holds the next data packet for the new wavelength for a guard time, as shown in Figure 3.11(c). The end-of-transmission packet is used to inform the destination node that the data transmission on the old wavelength has terminated and data will soon arrive on the new wavelength. The guard time prevents data from being lost during the transient period of lightpath migration.

3) After the guard time, the source node switches the data transmission from the old wavelength to the new wavelength. This is implemented by retuning the transmitter to the new wavelength, as shown in Figure 3.11(d).

4) Once the destination node receives the end-of-transmission packet, it switches data reception by retuning the wavelength of the receiver to the new wavelength, as shown in Figure 3.11(e) and Figure 3.11(f).

The guard time determines the disruption period of the rerouted lightpath. It is bounded by the sum of the following times:

- The switching time of the tunable optical transmitter at the source node
- The switching time of the tunable optical receiver at the destination node
- The processing time of the end-of-transmission packet
- The differential propagation time of sending packets through two wavelengths because of the wavelength dispersion of optical fibers or devices.

It was shown in [21] that, based on practical parameters, the total guard time is only of the order of microseconds, which is the only disruption period of the MTV-WR operation.

3.6.4 Lightpath Rerouting Schemes

The basic MTV-WR operation is used for rerouting a single lightpath. For multiple lightpaths, there are two basic rerouting schemes: parallel MTV-WR and sequential MTV-WR, both of which are based on the basic MTV-WR operation.

Parallel MTV-WR

In parallel MTV-WR [21], each of the rerouted lightpaths is moved to a free (or vacant) wavelength on the same route in parallel. For this purpose, the rerouted lightpath should be on a disjoint set of links. Because each lightpath migration is performed in parallel, the total delay for rerouting all the lightpaths is equal to the maximum of the delay for rerouting each lightpath. Accordingly, parallel MTV-WR has several advantages over other rerouting schemes, such as

- Shorter delay and disruption period
- Simpler implementation
- Less computational complexity

Example 3.10: Consider a route with five nodes and four links, and three wavelengths on each fiber link, as shown in Figure 3.12. Suppose that there are seven lightpaths already established and there is a new connection request for a lightpath between node 2 and node 4. Because there is no common wavelength on link (2, 3) and link (3, 4), the request cannot be directly accommodated. However, if path p_2 and path p_3 are moved to wavelength 2 and wavelength 3, respectively, wavelength 1 will become available on both link (2, 3) and link (3, 4). Therefore, path p_8 can be established on wavelength 1 for the connection request. In this example, path p_2 and path p_3 can be retuned to wavelength 2 and wavelength 3 simultaneously with the parallel MTV-WR scheme. If the delay for migrating p_2 is d_2, and the delay for migrating p_3 is d_3, the total delay of migrating the two lightpaths is equal to $\max(d_2, d_3)$.

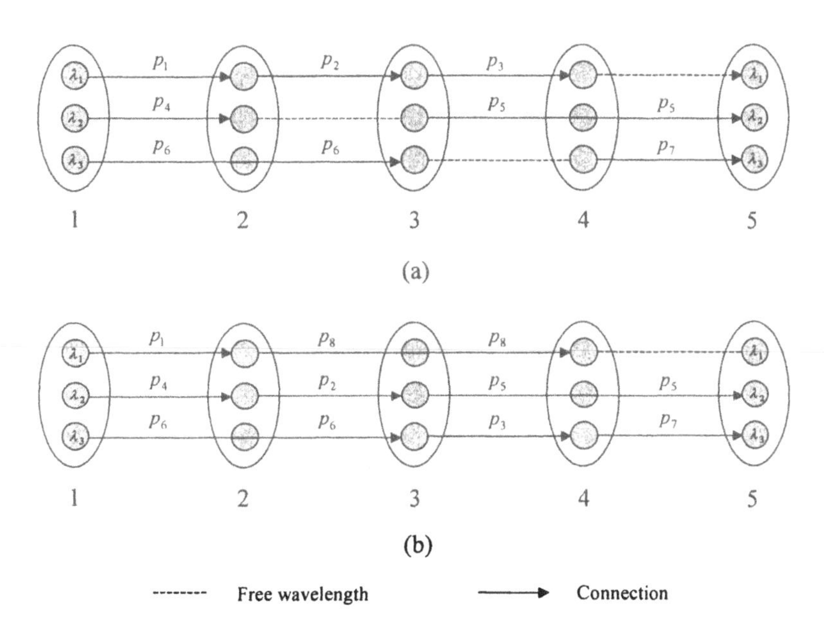

Figure 3.12 Illustration of parallel MTV-WR.

However, parallel MTV-WR cannot handle all situations. In many situations, the lightpaths to be rerouted cannot be moved in parallel. Instead, they can be migrated with the sequential MTV-WR scheme.

Sequential MTV-WR

In sequential MTV-WR [21], each of the rerouted lightpaths is moved to a free (or vacant) wavelength on the same route in a sequential manner. The total delay for rerouting all the lightpaths is the sum of the delay for rerouting each lightpath. Obviously, sequential MTV-WR largely increases the total rerouting delay. Despite this shortcoming, however, sequential MTV-WR is still a useful scheme because it can handle many situations that cannot be handled by parallel MTV-WR.

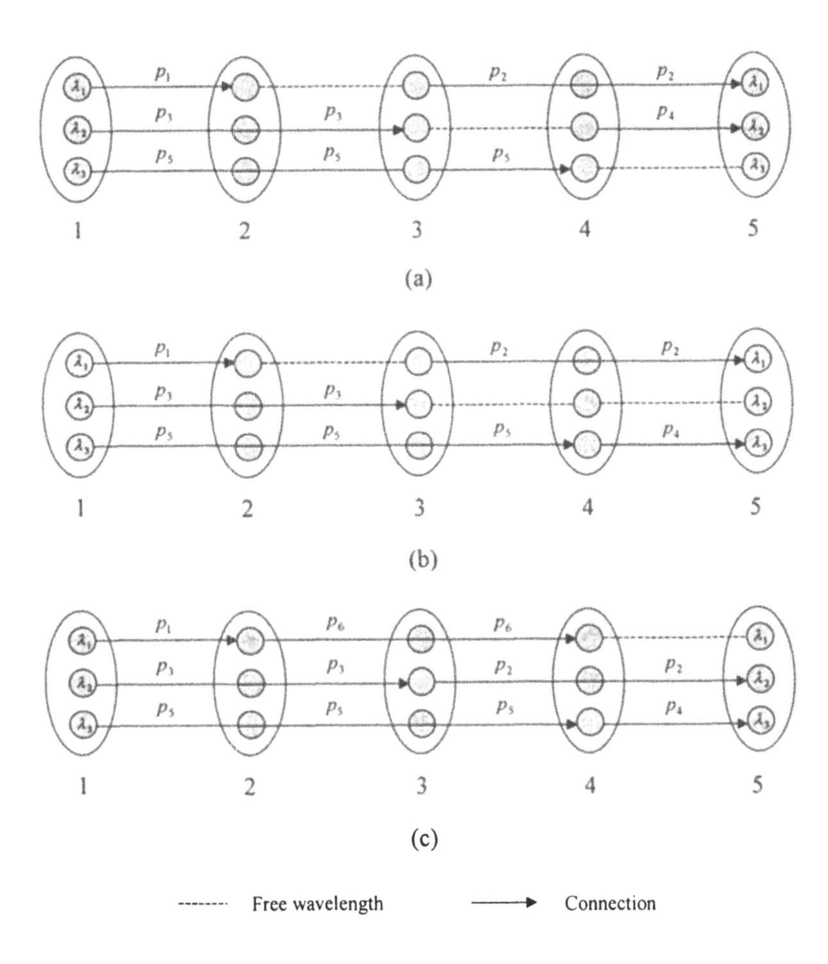

Figure 3.13 Illustration of sequential MTV-WR.

> ***Example 3.11***: Consider again the route in Example 3.10. Suppose that there are five lightpaths already established, as shown in Figure 3.13, and a new connection request arrives for a lightpath between node 2 and node 4. To accommodate this request, lightpath p_4 must first be retuned to wavelength 3, thus releasing wavelength 2. Then lightpath p_2 is retuned to wavelength 2, releasing wavelength 1 on both link (3, 4) and link (4, 5). Finally, path p_6 is established for the connection request with wavelength 1. In this example, the rerouted lightpaths, i.e., p_2, p_4, and p_6, do not have a disjoint set of links and thus cannot be migrated in parallel. If the delay for migrating p_2 is d_2, and the delay for migrating p_4 is d_4, the total delay of migrating the two lightpaths is equal to (d_2+d_4).

In addition to the situations that can be handled by either parallel or sequential MTV-WR, there are also some situations that cannot be handled by both parallel and sequential schemes. Such situations can be handled by a rerouting scheme called rerouting-after-stop (RAS).

Rerouting-After-Stop

In the rerouting-after-stop scheme, the network controller first sends control packets to the source nodes and destination nodes of the rerouted lightpaths to stop the data transmissions. In response, the source nodes stop the transmissions and send acknowledgments back to the controller. After receiving the acknowledgments from all the source nodes, the network controller sends a control packet to each routing node involved in the rerouting to change the state of the switch and set up a new route for each rerouted lightpath. After that, the routing nodes send acknowledgments back to the controller. After the controller receives the acknowledgments from all the routing nodes, it sends control packets again to the source nodes and destination nodes to restart the transmissions. Once a source node receives the control packet, it will resume the transmission on the corresponding new route. Therefore, the period of disruption with RAS is much longer than that with MTV-WR.

> ***Example 3.12***: Consider the situation shown in Figure 3.14. Suppose that there is a connection request for a lightpath between node 2 and node 4. To accommodate this request, lightpath p_2 must be retuned to wavelength 2 on link (3, 4) and link (4, 5) and lightpath p_4 must be retuned to wavelength 1 on link (4, 5). Because there is no free wavelength available to move the two lightpaths, the MTV-WR scheme cannot be used. However, if the transmissions on both

lightpaths are stopped, the two lightpaths can be then rerouted and lightpath p_6 can be thus established for the connection request.

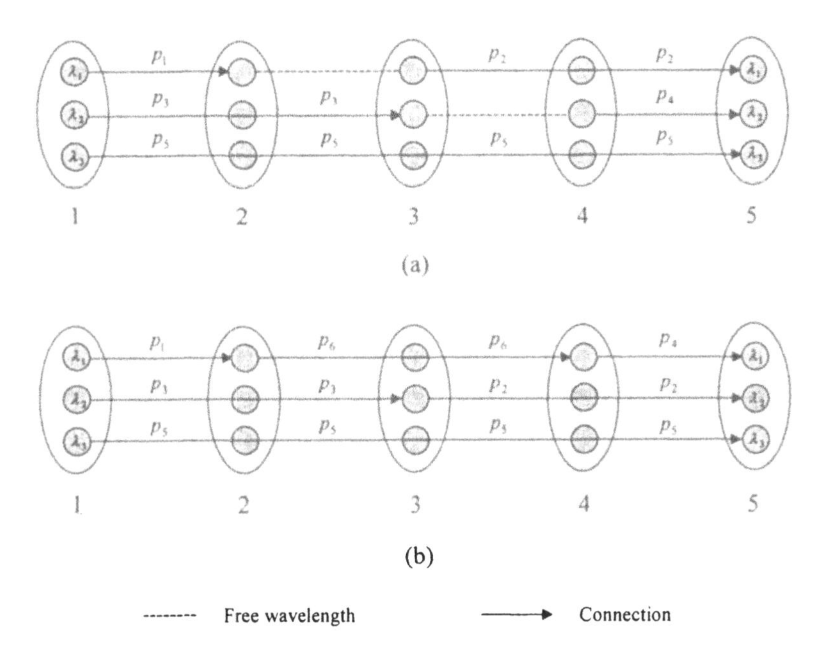

Figure 3.14 An example of a situation that the MTV-WR operation cannot handle.

3.6.5 Wavelength Rerouting Algorithms

A wavelength rerouting algorithm selects a set of rerouted lightpaths to create a wavelength-continuous route for a new connection request that otherwise will be blocked. As mentioned earlier, a rerouting algorithm must be simple in terms of computational complexity and be able to minimize the number of rerouted lightpaths. For this purpose, several rerouting algorithms have been proposed in the literature [21–24]. In this section, we present a well-known wavelength rerouting algorithm proposed in [21]. This rerouting algorithm attempts to minimize the weighted number of rerouted lightpaths in the network with the parallel MTV-WR scheme. For a selected route, the weighted number of rerouted lightpaths is defined as the weighted number of lightpaths whose links are traversed by the selected route. Each lightpath is counted only once no matter how many of its links are traversed by the selected route. The algorithm consists of three stages:

(1) auxiliary graph construction, (2) cost labeling, and (3) route searching, which are described as follows.

Stage 1: Auxiliary graph construction

The network is first represented as a graph *G(**V**, **E**)*, where ***V*** is a set of vertices representing the routing nodes in the network and ***E*** is a set of edges representing the links that connect the set of nodes. There are a set of *W* wavelengths on each fiber link. Assume that each node has a sufficient number of optical transmitters and receivers such that a connection request will not be blocked because of the unavailability of transmitters and receivers. To reduce the computational complexity of the rerouting algorithm, the graph is decomposed into *W* subgraphs, each corresponding to a wavelength. The vertices in a subgraph correspond to the routing nodes and the edges to the wavelength channels on the fiber links.

An auxiliary graph is then constructed by adding some crossover edges for each retunable lightpath to the graph. A lightpath is said to be retunable if there exists a wavelength, other than the one used by the lightpath, that is free on each link along the route of the lightpath. The crossover edges are used to count the weighted number of rerouted lightpahs to accommodate a new connection request. A crossover edge between two nodes for a retunable lightpath is created as long as there exist two or more edges on the lightpath between the two nodes. A retunable edge is either a crossover edge or an edge associated with a wavelength channel used by a retunable lightpath.

Stage 2: Cost labeling

A cost or weight is assigned to each edge in the auxiliary graph. The cost for a free edge is a tiny value, whereas the cost for a nonretunable edge is infinity. The cost for a retunable edge associated with a retunable lightpath is a positive value proportional to the hop count of the lightpath. The tiny value for a free edge is chosen such that the cost of any edge that is used by a retunable lightpath is larger than that of the longest route with only free edges.

Stage 3: Route searching

The objective of this algorithm is to find a route that minimizes the weighted number of rerouted lightpaths. The cost of a route is defined as the sum of the weights associated with all the edges that are traversed by the route, including those on the rerouted lightpaths and those free edges. For this purpose, a shortest-path routing algorithm is used to find the least-cost route on each of the *W* subgraphs and the route with the least cost value is selected to accommodate the connection request. If the least cost value is finite, the connection request will be accommodated. Otherwise, the connection request will be blocked. It should be pointed out that the least-cost route found traverses at most one edge or crossover edge of any rerouted ligthpath. This is because all the edges associated with a retunable lightpath are assigned the same cost. For this reason, the actual route corresponding to the least-cost route found is decided as follows.

- If the route traverses a free edge, the actual route should includc thc corresponding channel edge.
- If the route traverses a retunable edge of a retunable lightpath, the actual route should include those channel edges that are on the retunable lightpath and between the end vertices of the retunable edge.

The following example illustrates the three stages of the wavelength rerouting algorithm.

Example 3.13: Consider a simple network with five nodes and four links, as shown in Figure 3.15(a). Assume that each link has three wavelengths and there are three lightpaths already established in the network, which are p_1, p_2, and p_3. The network can be represented by a graph, as shown in Figure 3.15(b), in which there are three subgraphs, each corresponding to a wavelength. Each free edge is assigned a very tiny cost value of ε, such as edge (4, 5) in subgraph 1 and edges (1, 2), (2, 3), and (3, 4) in subgraph 2. Each nonretunable edge is assigned a cost value of ∞, such as edges (1, 2), (2, 3), (3, 4), and (4, 5) in subgraph 3. Each retunable edge associated with a retunable lightpath is assigned a positive cost value of c_i, which is proportional to the hop count of the lightpath. For example, edges (1, 2), (2, 3), and (3, 4) in subgraph 1, which are associated with a retunable lightpath p_1, are assigned 3. Edge (4, 5) in subgraph-2, which is associated with a retunable lightpath p_1, is assigned 1.

An auxiliary graph is then constructed by adding some crossover edges for each retunable lightpath to the graph, as shown in Figure 3.15(c). For example, there are three crossover edges created for lightpath p_1. Each of the crossover edges is assigned a cost value of 3. However, because lightpath p_2 consists of only one edge, no crossover edge is created for it.

Now suppose that there is a new connection request for a lightpath from node 1 to node 5. Obviously, the connection request cannot be accommodated directly without performing wavelength rerouting. To accommodate this connection request, a shortest-path routing algorithm is then used to find the least-cost route on each of the three subgraphs. Because the least costs on wavelength 1, wavelength 2, and wavelength 3 are $3+\varepsilon$, $1+3\varepsilon$, and ∞, respectively, the route on wavelength 2 is selected as it has the least cost among the three routes.

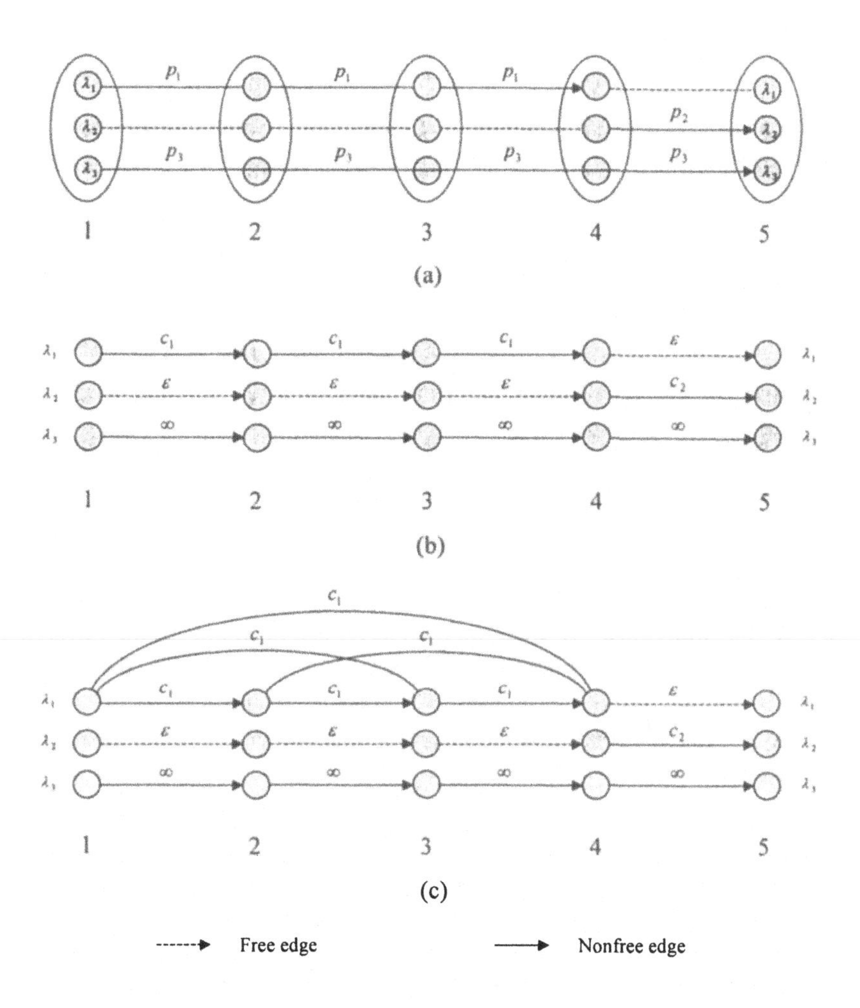

Figure 3.15 Illustration of graph construction: (a) network; (b) graph; (c) auxiliary graph.

3.7 Summary

In this chapter, we discussed the routing and wavelength assignment (RWA) problem in wavelength-routed WDM networks. The concept of the RWA problem was described, and both static RWA and dynamic RWA were discussed. The combined RWA problem can be formulated as an integer linear programming (ILP) problem, which is NP-complete. To make the problem more tractable, we can approximately divide the combined problem into a couple of subproblems, the routing subproblem and the wavelength assignment subproblem, and solve the two subproblems separately. For static traffic, the static routing subproblem can also be formulated as an ILP problem and the static wavelength assignment subproblem is equivalent to the graph-coloring problem. For dynamic traffic, there are a variety of routing algorithms and wavelength assignment algorithms already proposed in the literature. There are three basic types of routing paradigms: fixed routing, fixed-alternate routing, and adaptive routing. In general, fixed routing has the lowest computational complexity but may result in the largest blocking probability in the network. Adaptive routing can achieve the smallest blocking probability but has the highest computational complexity. Fixed-alternate routing provides a trade-off between computational complexity and blocking probability. These routing algorithms can be combined with different wavelength assignment algorithms to achieve desired network performance in different network scenarios. Fairness is also an important problem in routing and wavelength assignment. To improve fairness among short and long connections, an RWA algorithm must use some method to reduce or minimize the difference in terms of blocking probability for connection requests with different hop counts. Three different methods for improving fairness in RWA were briefly presented, including wavelength reservation, threshold protection, and limited alternate routing.

This chapter also discussed the wavelength rerouting problem. The need for wavelength rerouting was explained, and the problems involved in wavelength rerouting were described. Wavelength rerouting can effectively improve the wavelength utilization in the network and can also be used for handling network failures. However, it results in the disruptions of rerouted lightpaths and thus causes a significant amount of throughput loss in the network. For this reason, the incurred lightpath disruptions must be minimized during wavelength rerouting. Basic operations for lightpath migration were introduced, among which the MTV-WR operation combines the advantages of WR and MTV operations. Basic lightpath rerouting schemes were also introduced, among which parallel MTV-WR has several advantages over sequential MTV-WR, such as shorter disruption period and simpler implementation. In addition, a well-known wavelength rerouting algorithm was presented.

Problems

3.1 What are the main wavelength constraints in routing and wavelength assignment? What effects do they have on network performance?

3.2 Describe the concept of routing and wavelength assignment (RWA). What are the objectives of static RWA and dynamic RWA?

3.3 Give the advantages and disadvantages of centralized control and distributed control, respectively. Why is distributed control more suitable for large networks with dynamic traffic?

3.4 Why are heuristics needed to solve the RWA problem with static traffic?

3.5 To solve the RWA problem, why is the combined problem often divided into a routing subproblem and a wavelength assignment subproblem?

3.6 Give an example of static wavelength assignment using the graph-coloring method.

3.7 Qualitatively compare the performance and computational complexity of the three routing paradigms: fixed routing, fixed-alternate routing, and adaptive routing.

3.8 Compare a route that is selected by a fixed-alternate least-congested-path routing algorithm with one that is selected by a fixed-alternate routing algorithm and then apply the least-used wavelength assignment algorithm. Explain whether the two routes are the same and why.

3.9 What are fully adaptive routing and partially adaptive routing? Give an example of each of them.

3.10 What is the fairness problem in RWA? How can fairness be improved in RWA?

References

[1] R. Ramaswami and K. N. Sivarajan, "Routing and wavelength assignment in all-optical networks," *IEEE/ACM Transactions on Networking*, vol. 3, no. 5, Oct. 1995, pp. 489–500.

[2] S. Even, A. Itai, and A. Shamir, "On the complexity of timetable and multicommodity flow problems," *SIAM Journal of Computing*, vol. 5, 1976, pp. 691–703.

[3] Hui Zang, Jason P. Jue, and Biswanath Mukherjee, "A review of routing and wavelength assignment approaches for wavelength-routed optical WDM networks," *SPIE Optical Networks Magazine*, vol. 1, no. 1, Jan. 2000, pp. 47–60.

[4] D. Banerjee and B. Mukherjee, "A practical approach for routing and wavelength assignment in large wavelength-routed optical networks," *IEEE Journal on Selected Areas in Communications*, vol. 14, no. 5, Jun. 1996, pp. 903–908.

[5] M. R. Garey and D. S. Johnson, *Computer and Intractibility—A Guide to the Theory of NP Completeness*, W. H. Freeman, San Francisco, 1979.

[6] N. Biggs, "Some heuristics for graph colouring," in R. Nelson and R. J. Wilson, editors, *Graph Colorings*, Longman, New York, 1990, pp. 87–96.

[7] D. de Werra, "Heuristics for graph coloring," In G. Tinhofer, E. Mayr, and H. Noltemeier, editors, Computational Graph Theory, Volume 7 of Computing, Supplement, Springer-Verlag, Berlin, 1990, pp. 191–208.

[8] S. Subramaniam and R. A. Barry, "Wavelength assignment in fixed routing WDM networks," *Proceedings of IEEE ICC'97*, vol. 1, Montreal, Canada, Jun. 1997, pp. 406–410.

[9] G. Jeong and E. Ayanoglu, "Comparison of wavelength-interchanging and wavelength-selective cross-connects in multiwavelength all-optical networks," *Proceedings of IEEE INFOCOM'96*, vol. 1, San Francisco, Mar. 1996, pp. 156–163.

[10] E. Karasan and E. Ayanoglu, "Effects of wavelength routing and selection algorithms on wavelength conversion gain in WDM optical networks," *IEEE/ACM Transactions on Networking*, vol. 6, no. 2, Apr. 1998, pp. 186–196.

[11] R. A. Barry and S. Subramaniam, "The MAX-SUM wavelength assignment algorithm for WDM ring networks," *Proceedings of OFC'97*, Jun. 1997, pp. 121-122.

[12] X. Zhang and C. Qiao, "Wavelength assignment for dynamic traffic in multi-fiber WDM networks," *Proceedings of 7th International Conference on Computer Communications and Networks (IC3N'98)*, Lafayette, LA, Oct. 1998, pp. 479–485.

[13] A. Birman and A. Kershenbaum, "Routing and wavelength assignment methods in single-hop all-optical networks with blocking," *Proceedings of IEEE INFOCOM'95*, 1995, pp. 431–438.

[14] E. Dijkstra, "A note on two problems in connexion with graphs," *Numerische Mathematik*, vol. 1, 1959, pp. 269–271.

[15] R. Bellman, *Dynamic Programming*, Princeton University Press, Princeton, New Jersey, 1957.

[16] S. Ramamurthy and B. Mukherjee, "Fixed-alternate routing and wavelength conversion in wavelength-routed optical networks," *Proceedings of IEEE GLOBECOM'98*, vol. 4, Nov. 1998, pp. 2295–2302.

[17] H. Harai, M. Murata, and H. Miyahara, "Performance of alternate routing methods in all-optical switching networks," *Proceedings of IEEE INFOCOM'97*, vol. 2, Kobe, Japan, April 1997, pp. 516–524.

[18] K. Chan and T. P. Yum, "Analysis of least congested path routing in WDM lightwave networks," *Proceedings of IEEE INFOCOM'94*, vol. 2, Toronto, Canada, April 1994, pp. 962–969.

[19] L. Li and A. K. Somani, "Dynamic wavelength routing using congestion and neighborhood information," *IEEE/ACM Transactions on. Networking*, vol. 7, no. 5, Oct. 1999, pp.779–786.

[20] G. Mohan and C. Siva Ram Murthy, "Routing and wavelength assignment for establishing dependable connections in WDM networks," *Proceedings of IEEE International Symposium on Fault-Tolerant Computing*, Jun. 1999, pp. 94–101.

[21] Kuo-Chun Lee and Victor O. K. Li, "A wavelength rerouting algorithm in wide-area all-optical networks," *IEEE/OSA Journal of Lightwave Technology*, vol. 14, no. 6, Jun. 1996, pp. 1218–1229.

[22] G. Mohan and C. Siva Ram Murthy, "Efficient algorithms for wavelength rerouting in WDM multi-fiber unidirectional ring networks," *Computer Communications*, vol. 22, no. 3, Feb. 1999, pp. 232–243.

[23] G. Mohan and C. Siva Ram Murthy, "A time optimal wavelength rerouting algorithm for dynamic traffic in WDM networks," *IEEE/OSA Journal of Lightwave Technology*, vol. 17, no. 3, Mar. 1999, pp. 406–417.

[24] G. Mohan and C. Siva Ram Murthy, "Efficient wavelength rerouting in WDM single-fiber and multi-fiber networks with and without wavelength conversion," *Journal of High Speed Networks*, vol. 8, 1999, pp. 149–171.

Chapter 4

Virtual Topology Design

4.1 Introduction

An important problem in the design of wavelength-routed WDM networks is the virtual topology design problem [1–2]. A virtual topology consists of a set of all-optical connections or lightpaths established over the physical topology of a network to accommodate the traffic demand of network users. It forms an optical layer between the physical layer and the higher layers, such as SONET/SDH, ATM, and IP. This optical layer provides high-speed connectivity to its higher layers, which is protocol transparent and can thus support different kinds of network services at the higher layers. Because of the limitations in the number of wavelengths available on each fiber link as well as other network resources, such as transmitters and receivers, it may not be possible to establish a lightpath between each pair of nodes in the network. In this case, two nodes that cannot be connected directly by a lightpath have to use a concatenation of lightpaths to communicate. This would introduce electronic processing at each node that connects two consecutive lightpaths and thus affect network performance. For this reason, the virtual topology design problem becomes an issue of great concern. The objective of virtual topology design is to combine the advantages of optical transmission and electronic processing in order to enhance network characteristics and optimize network performance.

The design of a virtual topology is based on the traffic demand and physical topology of a network. However, such traffic demand and physical topology cannot be constant. There are many factors that may result in a change in traffic demand and physical topology. Under such changes, the virtual topology might become not optimal and therefore would need to be redesigned and reconfigured in response to the changing traffic demand or physical topology. The process for redesigning and reconfiguring a network from an old virtual topology to a new virtual topology is referred to as reconfiguration. The purpose of reconfiguration is to maintain the optimal network performance under various traffic and topology changes.

This chapter discusses the virtual topology design problem in wavelength-routed WDM networks. The basic virtual topology design problem is described, and the subproblems in virtual topology design are discussed. The basic concepts related to the virtual topology design are also introduced. The virtual topology design problem can be formulated as a mixed-integer linear (MILP) problem. A typical example of such formulations is presented with an objective to minimize the congestion in the network. Because an MILP problem is known to be NP hard, heuristics are often used to obtain approximate solutions to the problem. Several well-known heuristics for virtual topology design with regular topology, predetermined topology, and arbitrary topology are presented. To evaluate how exact or close an approximate solution is to an optimal solution, lower bounds on the network congestion and the number of wavelengths needed in the virtual topology design are discussed. Moreover, this chapter discusses the virtual topology reconfiguration problem. The need for virtual topology reconfiguration is explained, the main strategies for dealing with traffic changes and topology changes are introduced, and the basic methods for solving the reconfiguration problem are described. The focus of this chapter is primarily on the basic concepts and principles of virtual topology design for large transport networks with an arbitrary mesh physical topology.

4.2 Virtual Topology Design Problem

WDM technology is currently being deployed in large transport networks on a point-to-point transmission basis. In such point-to-point WDM networks, an optical signal is converted to an electronic signal for processing at each intermediate node and is then converted back to an optical signal for further transmission before it reaches the destination node. Although such networks have a huge transmission capacity, they may not have sufficient processing capacity to handle the traffic load arriving at each network node. This would lead to a severe performance bottleneck at each node. Moreover, because of the optical-electronic (O/E) and electronic-optical (E/O) conversion, the electronic processing cost at each node is largely increased because of the large amount of buffers and optical transponders required. All these factors impose a limitation on the delivery of the high bandwidth capacity to network users.

In a wavelength-routed WDM network, however, reconfigurable optical cross-connects (OXCs) are employed at each network node to provide the capability of switching optical signals entirely in the optical domain. By configuring such OXCs, the network is capable of establishing an all-optical connection called a lightpath between a pair of network nodes, which can deliver traffic entirely in the optical domain on an end-to-end basis without undergoing any O/E and E/O conversion. As a result, the traffic can avoid electronic buffering and processing at each intermediate node, which can thus alleviate the electronic bottleneck. A set of lightpaths that are established over the physical topology of a network constitutes a virtual topology of the network. Different virtual topologies can be established over the same physical topology. From the viewpoint of layered architecture, a virtual topology can be viewed as an optical layer between the physical layer and the higher layers of the network. This optical layer is protocol transparent and can support different kinds of network services at the higher layers, such as SONET/SDH, ATM, and IP. However, because of the limitations in the number of wavelengths available on each fiber link as well as other network resources like transmitters and receivers, it may not be possible to establish a lightpath between each pair of nodes in the network. In this case, two nodes that cannot be connected directly by a lightpath have to use a concatenation of lightpaths to deliver traffic. This would introduce electronic processing at each node that connects two consecutive lightpaths and thus affect network performance. For this reason, the virtual topology design problem becomes very important in achieving good network performance.

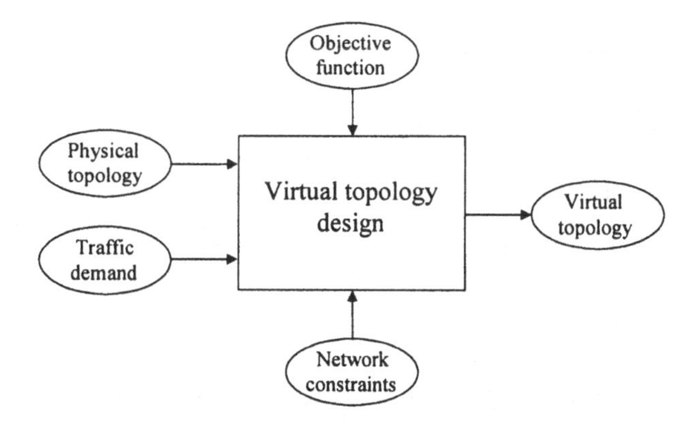

Figure 4.1 Illustration of virtual topology design problem.

Given the physical topology of a network, the constraints of network resources, and the traffic demand between each pair of network nodes, the problem of establishing a set of lightpaths over the physical topology to accommodate the traffic demand and meanwhile optimize the network performance is referred to as the virtual topology design problem, as illustrated in Figure 4.1. Network performance can be measured in terms of some performance metrics such as the number of hops between a pair of nodes, the offered traffic on a lightpath, the end-to-end delay from one node to another node, etc. In the following sections, we will first introduce the fundamentals of virtual topology design and then describe an exact formulation of the virtual topology design problem and present several well-known heuristics for solving the virtual topology design problem.

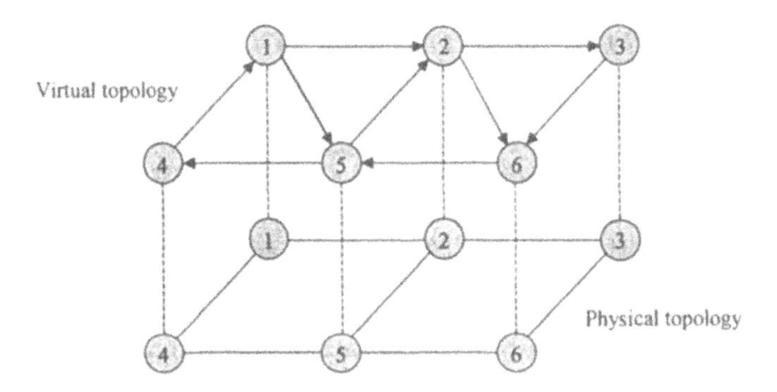

Figure 4.2 Illustration of physical and virtual topologies.

4.2.1 Physical and Virtual Topologies

The physical topology of a network consists of a set of network nodes that are interconnected by a set of fiber links. It can be represented by an undirected graph, in which each vertex corresponds to a node and each undirected edge corresponds to a pair of links between two nodes, each in one direction. The physical degree of a vertex in the graph is the number of undirected edges connected to the node. For example, the physical topology of a network with six nodes is illustrated in Figure 4.2. Node 1 has a degree of two, whereas node 5 has a degree of three. A virtual topology of the network, also called logic topology, consists of a set of logical links or lightpaths established over the physical topology. It can be represented by a directed graph, in which each vertex corresponds to a node in the physical topology and each directed edge corresponds to a logical link. The logical in-degree of a vertex is the number of incoming edges to the node. The logical out-degree of a vertex is the number of outgoing edges from the node. For example, a virtual topology over the physical topology is also shown in Figure 4.2. The logical in-degree and out-degree of node 1 are one and two, respectively. Both the logic in-degree and out-degree of node 5 are two.

A virtual topology is said to be symmetric if a logic link in one direction between two nodes always corresponds to another logic link in the other direction between the same two nodes, traversing the same set of intermediate nodes. Otherwise, it is said to be asymmetric. A virtual topology is said to be hop bound limited if the maximum number of physical hops that a logical link is allowed to traverse is limited. A virtual topology is said to be wavelength limited if the maximum number of wavelengths that can be used to establish a logical link is limited. The multiplicity of a virtual topology is defined as the maximum number of logic links between any pair of nodes, which is limited by the number of transmitters and receivers at each node. Such symmetry, wavelength, hop count, and multiplicity constraints may have a significant impact on the virtual topology design and therefore must be taken into account in the virtual topology design.

To achieve optimal network performance, it is desirable to establish a direct lightpath between each pair of network nodes in the physical topology, i.e., to design a fully connected virtual topology over the physical topology. However, this is usually not practical because of various limitations and constraints in the network. In general, the number of lightpaths that can be established in the network is limited by three factors:

- The number of wavelengths available on each fiber link
- The optical hardware resources available at each network node (e.g., transmitters and receivers)
- The electronic processing capacity at each network node

> ***Example 4.1***: Consider a network with N nodes. To establish a lightpath between each pair of nodes, the network must have a total number of $N(N-1)$ transceivers and a number of wavelengths to establish $N(N-1)$ lightpaths. Each node must have sufficient electronic processing capability to process the traffic load arriving on $(N-1)$ lightpaths simultaneously. For a smaller network with 10 nodes, it requires 90 transceivers. If each lightpath operates at 10 Gbps, each node must be capable of processing a traffic load at 90 Gbps. For a larger network with 100 nodes, it requires 9900 transceivers. Each node must be capable of processing 990 Gbps. This would result in a

huge amount of network cost and thus is economically impractical. Moreover, the number of wavelengths available on a single fiber link is still limited by enabling technology.

Because of such kinds of limitations and constraints, two nodes with no direct lightpath have to use a concatenation of lightpaths to communicate. At each intermediate node that connects two lightpaths, an optical signal received on one lightpath must be converted to its electronic form, switched electronically, and then converted back to its optical form before it is sent out on another lightpath to the destination node. Obviously, this would introduce longer delay in the delivery of traffic. However, this also provides a kind of flexibility for the delivery of traffic because it allows traffic to be dropped at an intermediate node. With this flexibility, traffic destined to the destination node can be converted back to its optical form and sent out onto the outgoing link while traffic destined to the local node can be delivered to the upper layer in its electronic form. Therefore, a virtual topology combines the advantages of optical transmission and electronic processing, and can thus achieve good network performance.

4.2.2 Subproblems in Virtual Topology Design

In general, the virtual topology design problem involves four aspects:

- Determining a set of lightpaths over the physical topology
- Routing each of the lightpaths
- Assigning a wavelength for each of the lightpaths
- Routing traffic over the lightpaths

Such a problem can be computationally difficult to handle. To make the problem tractable, the whole problem can be usually decomposed into four corresponding subproblems, topology design, lightpath routing, wavelength assignment, and traffic routing [3–9], which are then solved separately. It should be pointed out that such decomposition is approximate or inexact in the sense that solving the subproblems separately and combining the solutions may not obtain an optimal solution or even a solution for the whole problem. However, it provides a trade-off between optimal network performance and high computational complexity. The four subproblems are described in more detail as follows.

- Topology design: This subproblem is the determination of a set of lightpaths (or a virtual topology) over the physical topology only in terms of their source and destination nodes. In solving this subproblem, some resource constraints should be taken into account, such as the number of wavelengths available on each fiber link and the number of transceivers available at each network node.

- Lightpath routing: This subproblem is deciding on a physical route for each of the lightpaths determined in the topology design subproblem. In solving this subproblem, the number of hops on a physical route or the propagation delay on each fiber link should be taken into account. In some cases, the congestion of each fiber link should also be taken into account.
- Wavelength assignment: This subproblem is the assignment of an available wavelength for each of the lightpaths determined in the topology design subproblem. In solving this subproblem, the number of wavelengths available on each fiber link should be taken into account. In the absence of wavelength conversion, the wavelength-continuity constraint should also be taken into account.
- Traffic routing: This subproblem is the routing of the traffic over the virtual topology or the set of determined lightpaths in order to accommodate the traffic demand and achieve good network performance. This subproblem seems to be unessential to the virtual topology design problem because routing traffic over a virtual topology becomes relatively simple to implement as long as the other three subproblems are solved. In fact, there are many algorithms already proposed in the literature for solving the traffic routing subproblem. However, this subproblem should still be taken into account to make an exact formulation of the virtual topology design problem complete.

To better understand the four subproblems, let us examine the following example.

Example 4.2: Consider a physical topology illustrated in Figure 4.3(a). Assume that there are two transmitters and two receivers at each network node and two wavelengths on each fiber link. One solution to the topology design subproblem is illustrated in Figure 4.3(b). A lightpath is respectively determined between each pair of source and destination nodes as given in Column 1 of Table 4.1. One solution to the lightpath routing subproblem is given in Column 2 of Table 4.1. One solution to the wavelength assignment subproblem is given in Column 3 of Table 4.1. One solution to the traffic routing subproblem is given in Table 4.2.

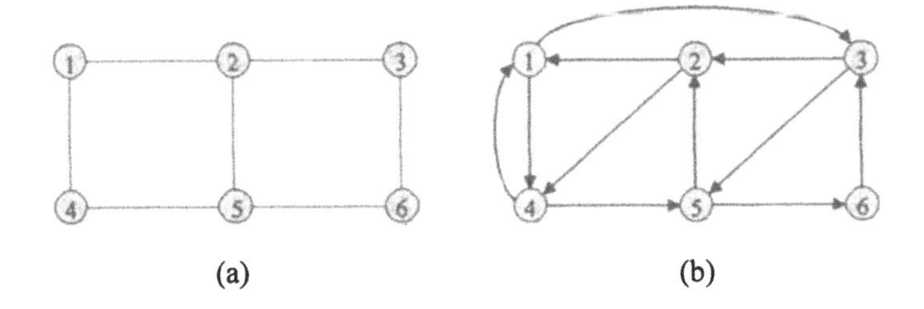

Figure 4.3 Illustration of four subproblems: (a) physical topology; (b) virtual topology.

Table 4.1

Lightpath	Route	Wavelength
(1, 3)	1-2-3	1
(1, 4)	1-4	1
(2, 1)	2-1	1
(2, 4)	2-1-4	2
(3, 2)	3-2	1
(3, 5)	3-2-5	2
(4, 1)	4-1	1
(4, 5)	4-5	1
(5, 2)	5-2	1
(5, 6)	5-6	1
(6, 3)	6-3	1

Table 4.2

(s, d)	Route	(s, d)	Route	(s, d)	Route
(1, 2)	1-3-2	(3, 1)	3-2-1	(5, 1)	5-2-1
(1, 3)	1-3	(3, 2)	3-2	(5, 2)	5-2
(1, 4)	1-4	(3, 4)	3-2-4	(5, 3)	5-6-3
(1, 5)	1-4-5	(3, 5)	3-5	(5, 4)	5-2-4
(1, 6)	1-4-5-6	(3, 6)	3-5-6	(5, 6)	5-6
(2, 1)	2-1	(4, 1)	4-1	(6, 1)	6-3-2-1
(2, 3)	2-1-3	(4, 2)	4-5-2	(6, 2)	6-3-2
(2, 4)	2-4	(4, 3)	4-1-3	(6, 3)	6-3
(2, 5)	2-4-5	(4, 5)	4-5	(6, 4)	6-3-2-4
(2, 6)	2-4-5-6	(4, 6)	4-5-6	(6, 5)	6-3-5

4.2.3 Virtual Topology Optimization

The virtual topology design problem can usually be formulated as a mixed-integer linear programming (MILP) problem with the objective of optimizing some network performance [2]. The important metrics used in the performance optimization can be the number of virtual hops traversed by a traffic unit, the offered traffic on a lightpath, or the end-to-end delay experienced by a traffic unit [3]. The optimization objectives based on such metrics are described as follows.

- Minimize the average weighted number of hops in the network: The average weighted number of virtual hops is defined as the average number of lightpaths traversed by one unit of traffic (e.g., one packet) in the network. Minimizing the average weighted number of hops in the virtual topology can result in a minimum number of O/E and E/O conversions and thus reduce the end-to-end delay experienced by the traffic. It can also increase the utilization of the network resources such as wavelengths and transceivers. Moreover, it can potentially reduce the offered traffic on a lightpath from different source and destination node pairs and thus reduce the queuing delay at an intermediate node.
- Minimize the congestion in the network: The congestion in the network is defined as the maximum offered traffic on any lightpath in the network. The offered traffic on a lightpath is defined as the aggregate traffic from all source and destination node pairs that use this lightpath. Minimizing the congestion in the network can balance the traffic load on each lightpath and thus accommodate more traffic in the network.
- Minimize the average end-to-end delay in the network: The end-to-end delay from a source node to a destination node includes the total propagation delay over all the lightpaths and the total processing delay at all the intermediate nodes from the source node to the destination node. The propagation delay over a lightpath is the total delay over all the physical links from one end to the other end of the lightpath. The processing delay at an intermediate node includes E/O conversion delay, queuing delay, and O/E conversion delay. In large transport networks, the propagation delay is generally larger than the processing delay. Minimizing the average end-to-end delay in the network can provide better quality of service to network users.

4.3 Virtual Topology Design Formulation

In this section, we present an exact formulation of the virtual topology design problem.

4.3.1 Terminology

We first define some terminology that is common to most formulations of the virtual topology design problem.

- Physical topology: An undirected graph that consists of a set of vertices and a set of edges. Each vertex corresponds to a physical node in the network, and each edge corresponds to a pair of fiber links between two nodes, each in one direction. Usually, there is a weight associated with each of the edges, which can be the link distance or the propagation delay over a link.
- Virtual topology: A directed graph that consists of a set of vertices and a set of edges. Each vertex corresponds to a vertex in the physical topology, and each edge corresponds to a logical link or lightpath. Usually, there is also a weight associated with each of the edges, which can be the lightpath distance or the propagation delay over a lightpath.
- Physical degree: The physical degree of a node is the number of physical links that directly connect the node to other nodes.
- Virtual degree: The virtual or logical degree of a node is the number of lightpaths that connect the node to other nodes.
- Virtual in-degree: The virtual in-degree of a node is the number of lightpaths that terminate at the node.
- Virtual out-degree: The virtual out-degree of a node is the number of lightpaths that originate at the node.
- Physical link propagation delay: The propagation delay over a physical link in the physical topology
- Logical link propagation delay: The propagation delay over a logical link or lightpath in the virtual topology, which is the overall propagation delay over all the physical links that the lightpath traverses
- Physical hop distance: The number of physical links that a lightpath traverses
- Logical hop distance: The number of logical links or lightpaths between a pair of nodes in the virtual topology

- Traffic matrix: A matrix that specifies the traffic demand between each pair of nodes in the physical topology. The traffic demand can be given in terms of the number of packets arriving per second or the amount of bandwidth required. This matrix describes the traffic pattern of the network.
- Hop matrix: A matrix that specifies the maximum number of physical hops that a lightpath between a pair of nodes is allowed to traverse
- Link indicator: An indicator used to indicate whether there is a physical link between a pair of nodes in the physical topology
- Lightpath indicator: an indicator used to indicate if there is a logic link or lightpath between a pair of nodes in the virtual topology
- Virtual traffic load: The virtual traffic load over a lightpath is the aggregate traffic offered to that lightpath, which is the aggregate traffic from each source and destination node pair that uses the lightpath.

4.3.2 Formulation

In the following formulation, we consider an objective function that minimizes the congestion of the network. The notations used in the formulation are defined as follows.

Subscripts

- s, d: the source and destination nodes of an end-to-end connection
- i, j: the end nodes of a logical link or lightpath
- m,n: the end nodes of a physical link
- k: the kth wavelength

Parameters

- N: the number of network nodes in the network
- W: the number of wavelengths available on each fiber link
- T: an $N \times N$ traffic matrix defined as follows

$$\boldsymbol{T}=[\lambda_{sd}]$$

where λ_{sd} denotes the aggregate traffic from source node s to destination node d

- H: an $N \times N$ hop matrix defined as follows

$$\boldsymbol{H}=[h_{ij}]$$

where h_{ij} denotes the maximum number of physical hops that a lightpath is allowed to traverse from node i to node j

- p_{mn}: a link indicator defined as follows

$$p_{mn} = \begin{cases} 0 & \textit{no physical link exists from node } m \textit{ to node } n \\ 1 & \textit{otherwise} \end{cases}$$

- Δ_{in}: the virtual in-degree of a node
- Δ_{out}: the virtual out-degree of a node

Variables

- λ_{ij}: the traffic load offered to logic link (i, j) from all source and destination node pairs
- λ_{ij}^{sd}: the traffic load offered to logic link (i, j) from source and destination node pair (s, d)
- b_{ij}: a lightpath indicator defined as follows

$$b_{ij} = \begin{cases} 0 & \textit{no lightpath exists from node } i \textit{ to node } j \\ 1 & \textit{otherwise} \end{cases}$$

- $c_{ij}(k)$: a lightpath-wavelength indicator defined as follows

$$c_{ij}(k) = \begin{cases} 1 & \textit{a lightpath from node } i \textit{ to} \\ & \textit{node } j \textit{ uses wavelength } k \\ 0 & \textit{otherwise} \end{cases}$$

- $c_{ij}^{mn}(k)$: a lightpath-wavelength-link indicator defined as follows

$$c_{ij}^{mn}(k) = \begin{cases} 1 & \textit{a lightpath from node } i \textit{ to node } j \\ & \textit{uses wavelength } k \textit{ and traverses} \\ & \textit{a link from node } m \textit{ to node } n \\ 0 & \textit{otherwise} \end{cases}$$

Objective

$$Min[\lambda_{\max}] \tag{4-1}$$

where $\lambda_{\max} = \max(\lambda_{ij}) \quad \forall i, \forall j$

Degree constraints

$$\sum_{i} b_{ij} = \Delta_{in} \qquad (4\text{-}2)$$

$$\sum_{j} b_{ij} = \Delta_{out} \qquad (4\text{-}3)$$

Constraints (4-2) and (4-3) ensure that the virtual topology is constrained by the virtual in-degree and out-degree of a node. Note that it has been assumed that each node has the same in-degree and out-degree, respectively. The virtual in-degree and out degree of a node are constrained by the number of receivers and transmitters available at the node, respectively.

Traffic constraints

$$\lambda_{ij} \le \lambda_{\max} \qquad \forall(i,j) \qquad (4\text{-}4)$$

$$\lambda_{ij} = \sum_{sd} \lambda_{ij}^{sd} \qquad \forall(i,j) \qquad (4\text{-}5)$$

$$\lambda_{ij}^{sd} \le b_{ij}\lambda_{sd} \qquad \forall(i,j), \forall(s,d) \qquad (4\text{-}6)$$

$$\sum_{j} \lambda_{ij}^{sd} - \sum_{j} \lambda_{ji}^{sd} = \begin{cases} \lambda_{sd} & i = s \\ -\lambda_{sd} & i = d \\ 0 & otherwise \end{cases} \qquad \forall(s,d)\ \forall i \qquad (4\text{-}7)$$

Constraint (4-4) defines the congestion of the network, that is, the maximum offered traffic on any lightpath in the network. Constraint (4-5) specifies that the total amount of traffic offered to a lightpath is the overall traffic offered to the lightpath from all source and destination node pairs in the network. Constraint (4-6) captures the fact that the traffic offered to a lightpath from a particular source and destination node pair cannot be more than the overall traffic for that source and destination node pair. Constraint (4-7) ensures the conservation of traffic flows at the end nodes of a lightpath.

Wavelength constraints

$$\sum_{k=0}^{W-1} c_{ij}(k) = b_{ij} \qquad \forall(i,j) \qquad (4\text{-}8)$$

$$c_{ij}^{mn}(k) \le c_{ij}(k) \qquad \forall(i,j), \forall(m,n), \forall k \qquad (4\text{-}9)$$

$$\sum_{ij} c_{ij}^{mn}(k) \le 1 \qquad \forall (m,n), \forall k \tag{4-10}$$

$$\sum_{k=0}^{W-1}\sum_{m} c_{ij}^{mn}(k) p_{mn} - \sum_{k=0}^{W-1}\sum_{m} c_{ij}^{nm}(k) p_{nm} = \begin{cases} b_{ij} & n = j \\ -b_{ij} & n = i \\ 0 & otherwise \end{cases} \tag{4-11}$$

$$\forall (i,j), \forall n$$

Constraint (4-8) ensures that if a lightpath exists between a pair of nodes, the wavelength used by the lightpath is unique. Constraint (4-9) ensures the consistency of wavelength assignment. Constraint (4-10) ensures that a wavelength can be used by at most one lightpath on each physical link, that is, two lightpaths cannot use the same wavelength on each physical link. Constraint (4-11) ensures the conservation of wavelengths at the end nodes of each physical link of a lightpath.

Hop constraints

$$\sum_{mn} c_{ij}^{mn}(k) \le h_{ij} \qquad \forall (i,j), \forall k \tag{4-12}$$

Constraint (4-12) ensures the number of physical links that a lightpath is allowed to traverse.

The inputs to the formulation are the given parameters, including the physical topology p_{mn}, the traffic matrix $\boldsymbol{T}$, the hop matrix $\boldsymbol{H}$, the number of wavelengths available on each fiber link W, and the logical degree of each node Δ_{in} :and Δ_{out}. The outputs from the formulation are the variables that relate to the virtual topology, lightpath routing, wavelength assignment, and traffic routing over the virtual topology. The lightpath indicators b_{ij} provide a virtual topology or a set of lightpaths in terms of their source and destination nodes. The lightpath-wavelength indicators $c_{ij}(k)$ and lightpath-wavelength-link indicators $c_{ij}^{mn}(k)$ provide the physical links that each lightpath consists of and the wavelength assigned to each lightpath. The virtual traffic loads λ_{ij} and λ_{ij}^{sd} provide the routing of the traffic between each source and destination node pair over the virtual topology.

It should be pointed out that this formulation is just a specific example for solving the virtual topology design problem. In the literature, we can find many other formulations [4–10], which consider different objectives and constraints and address all or some of the subproblems. For example, the objective of the exact formulation in [4] is to minimize the average packet delay. The exact formulation in [6] does not consider the number of wavelengths available on a single fiber link and thus does not address the wavelength assignment subproblem. The physical hop bound is not considered in the virtual topology, either. Instead, there is a constraint on the average delay between each source and destination node pair.

In general, a mathematical program is called a linear program (LP) if all the objective functions and constraint equations are linear and all the variables take real values. It is further called an integer linear program (ILP) if all the variables take integer values. If only some of the variables take integers, the program is called a mixed-integer linear program (MILP). The program formulated above is an example of an MILP because variables b_{ij} take integer values whereas variables λ_{ij} and λ_{ij}^{sd} may not. A solution to a mathematical program is any set of the variable values that are subject to all the constraints. An optimal solution is a solution that optimizes (either maximizes or minimizes) an objective function. The value of a mathematical program is the value of the objective function in the optimal solution. Although there are many efficient algorithms for solving LPs, no efficient algorithms are available for solving arbitrary ILPs and MILPs [1].

4.4 Heuristics

The virtual topology design problem can be formulated as a mixed-integer linear program (MILP). This program has been proved to be a NP-hard problem [12] and thus is computationally intractable. In fact, it is not practical to obtain an exact solution to the formulated problem for networks of larger sizes. To address this problem, the whole problem can be approximately decomposed into four subproblems, as described in Section 4.2.2. A solution to the whole problem depends on the solutions to the subproblems. However, some of the subproblems are also known to be NP hard [4][12], such as the topology design subproblem. For this reason, heuristics are often used to obtain approximate solutions to the subproblems or the whole problem. In this section, we first discuss some lower bounds that are used to evaluate a heuristic solution and then present several well-known heuristic algorithms that have been proposed in the literature.

4.4.1 Bounds

Heuristics can only provide approximate solutions to the virtual topology design problem. To evaluate how exact or close an approximate solution is to an optimal solution, it is necessary to know the lower bounds on the optimal solution, which are derived theoretically. In this section, we discuss such lower bounds on the network congestion and the number of wavelengths needed in the virtual topology design.

Lower bounds on the network congestion

The network congestion is often used as an optimization metric in the virtual topology design. A lower bound on the network congestion indicates that any solution cannot achieve a smaller value of congestion than the lower bound. We now discuss three lower bounds on the congestion, which are derived in [6].

Let $\overline{H}$ denote the traffic-weighted average number of virtual hops between a pair of source and destination nodes in the virtual topology, E_l denote the number of lightpaths in the virtual topology, and λ denote the total arrival rate of packets to the network. It is easy to see that

$$\lambda_{\max} \geq \lambda \overline{H} / E_l \tag{4-13}$$

Hence a lower bound on $\overline{H}$ gives a lower bound on the network congestion. To derive a lower bound on $\overline{H}$, we can use the result obtained in [13] as follows. For any virtual topology with N nodes and a maximum degree of Δ_l, there are at most $N\Delta_l$ source and destination node pairs that can be connected by one hop, $N\Delta_l^2$ pairs by two hops, $N\Delta_l^3$ pairs by three hops, and so on. To minimize the traffic-weighted number of virtual hops, the node pairs with the largest amount of traffic should be connected by the smallest number of logical hops. Accordingly, consider a topology in which the $N\Delta_l$ node pairs with the largest amount of traffic are connected by one-hop paths, the $N\Delta_l^2$ node pairs with the next largest amount of traffic are connected by two-hop paths, and so on. As a result, the traffic-weighted average number of virtual hops in such a topology becomes a lower bound, that is,

$$\overline{H} \geq \sum_k kS_k \tag{4-14}$$

where S_k denotes the total amount of traffic between the node pairs with the number of hops being k.

The bound given by (4-13) is based on the fact that if the total amount of traffic to the network were equally distributed among all the lightpaths in the network, the traffic on each logical link would be the congestion of the network. The value of this congestion can be used as a lower bound for any virtual topology designed for the network under the same traffic conditions. This bound takes into account the traffic demand, but not the traffic pattern. It assumes that the traffic between any pair of source and destination nodes can be distributed to any lightpath in the virtual topology. For this reason, it is referred to as the physical topology-independent bound in [2].

Based on the physical topology-independent bound, a stronger lower bound on the network congestion is derived in [6], which is referred to as the minimum flow tree bound. This bound is actually based on a stronger lower bound on $\overline{H}$, which is obtained by focusing the consideration on each source node rather than the entire network. For each source node, there are at most Δ_l destinations that can be connected to the source node by one hop, Δ_l^2 destinations by two hops, and so on. Accordingly, consider a topology in which for each source node, the destinations with the largest amount of traffic are connected by one-hop paths, the destinations with the next largest amount of traffic are connected by two-hop paths, and so on. As a result, the traffic-weighted average number of virtual hops in such a topology is a stronger lower bound on $\overline{H}$ and we have

$$\overline{H} \geq \overline{H}_{\min} \tag{4-15}$$

and the minimum flow tree bound can be expressed as

$$\lambda_{\max} \geq \lambda \overline{H}_{\min} E_l \tag{4-16}$$

The expression of $\overline{H}_{\min}$ as well as its derivation can be found in [6].

Based on the minimum flow tree bound, a much stronger lower bound is further derived in [6], which is referred to as the iterative bound. This bound is derived by aggregating and relaxing the MILP formulation. Because the MILP formulation in Section 4.3 considers each source and destination node pair as a commodity, it is usually referred to as a disaggregate formulation [14]. By considering each source node, rather than each source and destination node pair, as a commodity, a more tractable MILP formulation can be obtained, which is referred to as an aggregate MILP formulation. The additional constraint added to the aggregate MILP formulation [14] is

$$\lambda_{\max} \geq \sum_{s} \lambda_{ij}^{s} + \lambda_{\max}^{L} (1 - b_{ij}) \tag{4-17}$$

where $\lambda_{\max}^{L}$ is any lower bound known a priori, e.g., the minimum flow trcc bound, λ_{ij}^{s} is the traffic load offered on logical link (i, j) from source node s, and b_{ij} is the lightpath indicator defined in Section 4.3. By relaxing the constraints $b_{ij} \in \{0,1\}$ to $0 \le b_{ij} \le 1$, the aggregate MILP formulation reduces to a linear program (LP). The added constraint is superfluous in the MILP formulation but not in the relaxed LP formulation. By solving the LP, an initial lower bound $\lambda_{\max}^{L}(0)$ (e.g., the minimum flow tree bound) can lead to another lower bound $\lambda_{\max}^{L}(1)$. To improve the lower bound, we can use an improved lower bound $\lambda_{\max}^{L}(k)$, $k \ge 1$ to obtain a further improved lower bound $\lambda_{\max}^{L}(k+1)$ on an iterative basis. Such lower bounds are referred to as the iterative bounds. It has been pointed out in [6] that for the NFSNET network a lower bound obtained by 25 iterations cannot be significantly improved by further iterations.

Lower bounds on the number of wavelengths

In the virtual topology design, it is desirable to use as few wavelengths as possible because of the limitation in the number of wavelengths available on each fiber link. This limitation may be considered in an exact formulation of the virtual topology design problem as a wavelength constraint as in the formulation of Section 4.3. However, it is often not considered in a heuristic algorithm in order to make the problem computationally tractable. For this reason, a lower bound on the number of wavelengths needed for a particular problem is very important for evaluating the solution found by a heuristic algorithm. We now discuss two lower bounds on the number of wavelengths, which are also derived in [6].

Consider a node with the minimum physical degree of Δ_p in the physical topology. To design a virtual topology with the maximum logical degree of Δ_l, this node must route Δ_l lightpaths over Δ_l physical links. Accordingly, the average number of wavelengths routed over one of the physical links is Δ_l / Δ_p, that is, the lower bound on the number of wavelengths is

$$W \ge \Delta_l / \Delta_p \tag{4-18}$$

where W denotes the number of wavelengths needed on each fiber link. This bound is referred to as the physical topology degree bound in [2].

To have a better bound on W, consider that a physical topology with N nodes and E_p undirected physical links has $2E_p$ directed physical links. Each directed physical link can be traversed by multiple lightpaths in each direction. Accordingly, the average number of lightpaths traversing a directed physical link can be obtained by dividing the total number of directed physical links traversed by all the lightpaths by the number of directed physical links in the physical topology. Let h_{ij} denote the number of directed physical links on the shortest path from node i to node j. Let $E_i(\Delta_l)$ denote the total number of directed physical links on the Δ_l shortest paths from node i. Hence the total number of directed physical links traversed by the Δ_l lightpaths from node i is at least $E_i(\Delta_l)$. Therefore, the total number of directed physical links traversed by the $N\Delta_l$ lightpaths in the virtual topology is at least $\sum E_i(\Delta_l)$. The average number of lightpaths traversing a directed physical link gives a lower bound on W as follows, which is referred to as the physical topology link bound in [2].

$$W \geq [\sum_i E_i(\Delta_l)]/2E_p \tag{4-19}$$

In addition to the lower bounds discussed above on the network congestion and the number of wavelengths, there are also other lower bounds derived in the literature. The readers are referred to [12] and [15] for more details.

4.4.2 Heuristics Design

As mentioned earlier, heuristics can be used to obtain approximate solutions to the virtual topology design problem or its subproblems. A variety of heuristic algorithms are proposed in the literature. Most of these heuristics address only some rather than all of the subproblems. In some heuristics, an assumption is made about the virtual topology, which can be a regular topology or an irregular topology. Accordingly, such heuristics only need to address the lightpath routing and wavelength assignment subproblems. In other heuristics, no particular assumption is made about the virtual topology. These heuristic algorithms must address the topology design subproblem as well as some or all of the other subproblems.

Design with regular topology

In the heuristic design, regular topologies can be a choice for the virtual topology to be embedded over a physical topology. A topology is regular in the sense that all the nodes in the topology have the same degree. Typical examples of regular topologies are hypercubes and shufflenets [1], as shown in Figure 4.4. Regular topologies have some advantages over irregular topologies. For example, they are simple in traffic routing and tolerant to network faults. They can provide good performance (e.g., high throughputs and low delays) under uniform traffic patterns [2–3]. However, if the traffic pattern is not uniform, a regular topology may not be able to provide good performance. In regular virtual topology design, all four subproblems need to be addressed. In particular, the topology design subproblem must consider the choice of a regular topology as well as the node mapping between the physical topology and the virtual topology, which is referred to as the node-mapping subproblem. Once a regular topology is chosen, each of the nodes in the physical topology must be mapped to a node in the regular topology. The node-mapping subproblem has been known to be NP hard [2]. Accordingly, heuristics are needed to solve this subproblem. The node-mapping subproblem is unique to the regular virtual topology design. It does not exist in predetermined or arbitrary virtual topology design. However, the traffic routing subproblem is simplified because of the choice of a regular topology. In Section 4.4.3, we will present two well-known heuristic algorithms proposed for regular virtual topology design in the literature.

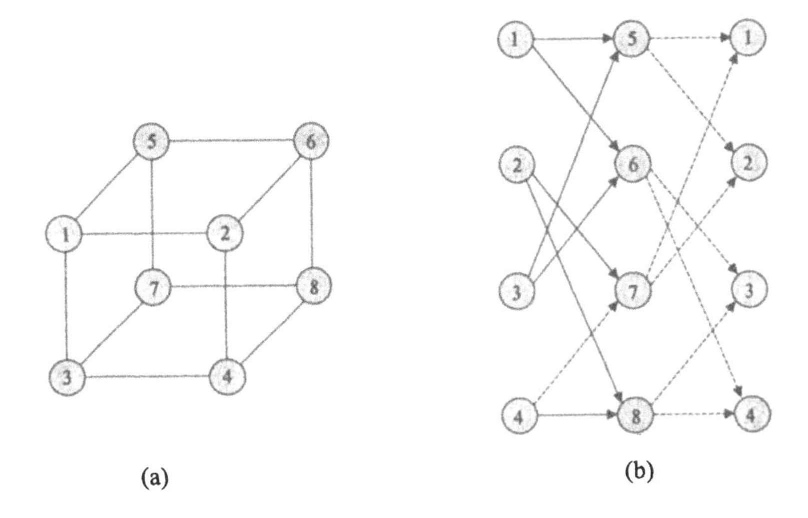

Figure 4.4 Regular topologies: (a) 8-node hypercube; (b) 8-node shufflenet.

Design with predetermined topology

In some heuristic designs, the virtual topology in terms of a set of lightpaths with their source and destination nodes is supposed to be known a priori or predetermined. A heuristic algorithm only needs to solve the subsequent three subproblems, in particular the lightpath routing and wavelength assignment subproblems. In this case, the traffic pattern in the network is usually taken into account in the predetermined virtual topology. The focus is on the lightpath routing and wavelength assignment subproblems. In Section 4.4.4, we will present two well-known heuristic algorithms proposed for predetermined virtual topology design in the literature.

Design with arbitrary topology

In other heuristic designs, no assumption is made on the virtual topology to be embedded on a physical topology. A heuristic algorithm must solve the topology design subproblem as well as some or all of the subsequent subproblems. In this case, the traffic pattern in the network is usually taken into account in the topology design subproblem. An arbitrary topology is required to address nonuniform traffic patterns and irregular physical topologies. In Section 4.4.5, we will present three well-known heuristic algorithms proposed in the literature for arbitrary virtual topology design.

4.4.3 Heuristics for Regular Topology Design

In this section, we present two heuristics for solving the virtual topology design problem with regular topologies. Node mapping is a unique subproblem in regular virtual topology design that does not exist in irregular virtual topology design. The other subproblems are the same as those discussed in Section 4.2.2. The following example illustrates the node-mapping subproblem.

> ***Example 4.3***: Consider a physical network with eight nodes as shown in Figure 4.5(a). Assume that a hypercube is chosen to be the virtual topology. A solution to the node-mapping subproblem is shown in Figure 4.5(b).

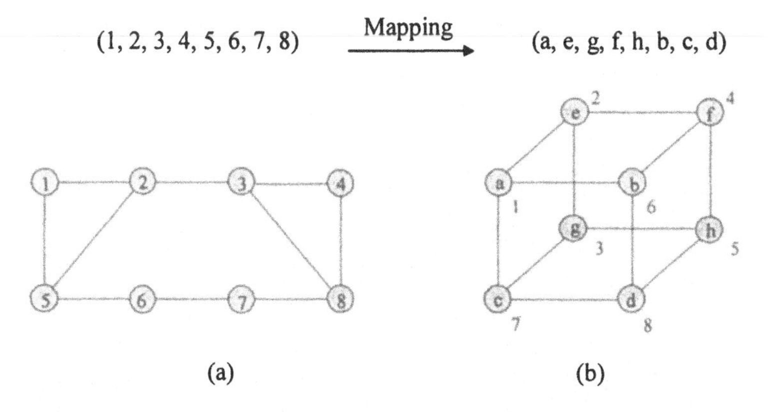

Figure 4.5 Illustration of node mapping: (a) physical topology; (b) hypercube virtual topology.

Greedy-Based Heuristic

A heuristic algorithm based on a greedy algorithm for node mapping with a hypercube has been proposed in [4]. The objective of this heuristic is to minimize the total average propagation delay in the virtual topology. The heuristic algorithm can be briefly described as follows.

- The node-mapping process starts with an initial node mapping, which is obtained by mapping the nodes in descending order of their physical degrees and meanwhile mapping adjacent nodes in the physical topology to adjacent nodes in the virtual topology as far as possible. This initial mapping does not necessarily have to result in the minimum propagation delay.
- The initial mapping is then refined in an iterative and greedy manner by handling each node in the reverse order of the initial node mapping. Each node is swapped with another node if this can reduce the average propagation delay. After all the nodes have been handled, the node mapping process is completed and the topology design subproblem is solved.
- For the lightpath routing subproblem, each lightpath in the virtual topology is routed over the shortest path in the physical topology.
- For the wavelength assignment subproblem, the physical links are handled in descending order of the number of lightpaths that traverse the links. For each physical link, each lightpath is assigned a wavelength if the lightpath has not been assigned a wavelength on some other physical link. If a sufficient number of wavelengths are available on each physical link, this heuristic algorithm can provide a good solution to the virtual topology design problem. Otherwise, the solution is not guaranteed.
- For the traffic routing subproblem, the shortest path is also used to route the traffic between a pair of nodes.

To better understand the heuristic algorithm, let us examine the following example.

> ***Example 4.4***: Consider a physical topology that has six nodes and nine links with a sufficient number of wavelengths available on each physical link, as shown in Figure 4.6(a). Assume that a hypercube topology with eight nodes is chosen to be the virtual topology, as shown in Figure 4.6(b). The traffic matrix is as follows.

$$\begin{bmatrix} 0 & 0.50 & 0.00 & 0.35 & 0.00 & 0.60 \\ 0.00 & 0 & 0.25 & 0.00 & 0.95 & 0.00 \\ 0.45 & 0.00 & 0 & 0.50 & 0.00 & 0.65 \\ 0.00 & 0.15 & 0.00 & 0 & 0.35 & 0.00 \\ 0.20 & 0.00 & 0.65 & 0.00 & 0 & 0.80 \\ 0.00 & 0.60 & 0.00 & 0.85 & 0.00 & 0 \end{bmatrix}$$

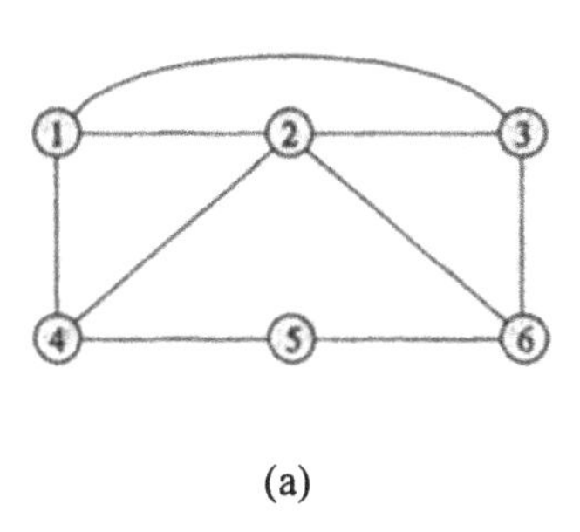

(a)

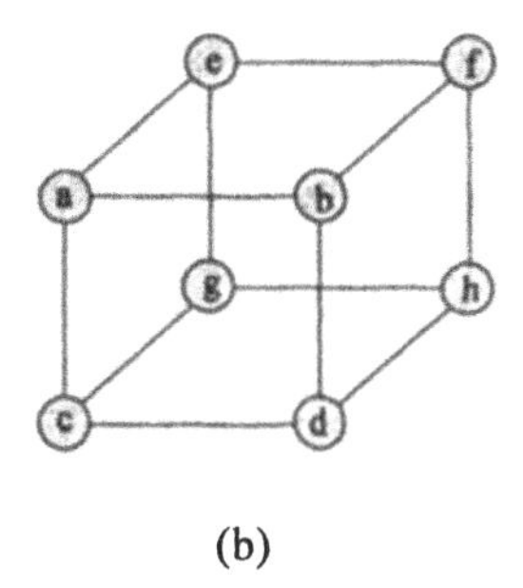

(b)

Figure 4.6 Illustration of greedy-based heuristic (1): (a) physical topology; (b) hypercube virtual topology.

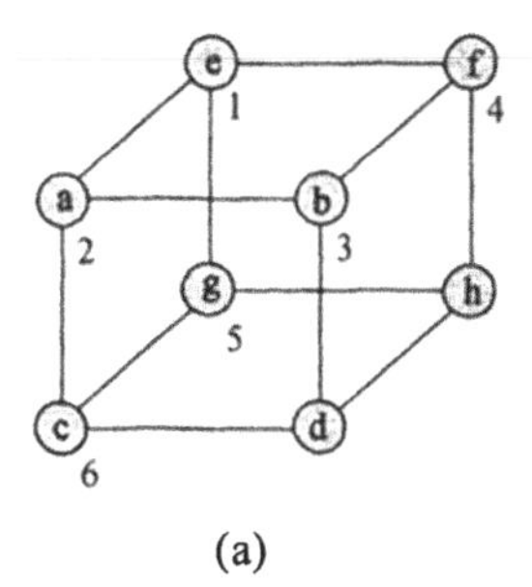

(a)

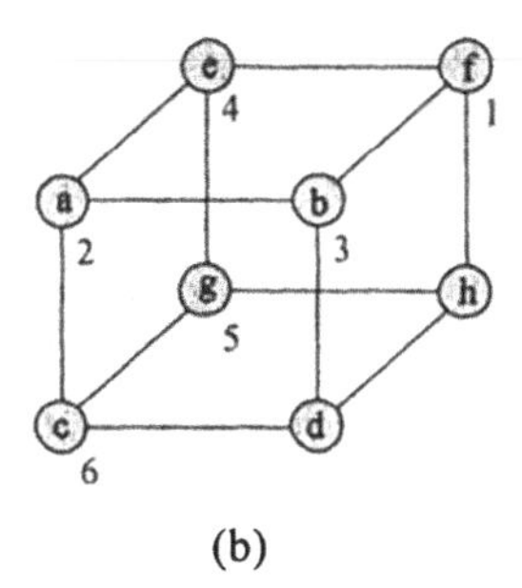

(b)

Figure 4.7 Illustration of greedy-based heuristic (2): (a) an initial mapping; (b) a refined mapping.

According to the described heuristic, an initial node mapping is first obtained, as shown in Figure 4.7(a). Table 4.3 shows the corresponding virtual and physical routes as well as the propagation delay between each pair of nodes with nonzero entry in the traffic matrix. In this case, the average propagation delay is 28/15. Note that each lightpath in the virtual topology is routed over the shortest path in the physical topology. The initial mapping is then refined in an iterative manner by handling each node in the reverse order of the initial node mapping. For example, because a swapping of node 1 and node 4 in the initial mapping can result in a reduced average propagation delay, a new mapping can thus be obtained, as shown in Figure 4.7(b) and Table 4.4. In this case, the average propagation delay becomes 25/15. Suppose that after handling all the nodes, the virtual topology obtained is shown in Figure 4.7(b). A solution to traffic routing over the virtual topology has already been shown in Column 2 and Column 3 of Table 4.4. Note that because there are a sufficient number of wavelengths available on each physical link, wavelength assignment is considered in this example.

Table 4.3

Node pair	Virtual route	Physical route	Number of hops
(1, 2)	1-2	1-2	1
(1, 4)	1-4	1-4	1
(1, 6)	1-2-6	1-2-6	2
(2, 3)	2-3	2-3	1
(2, 5)	2-6-5	2-6-5	2
(3, 1)	3-2-1	3-2-1	2
(3, 4)	3-4	3-2-4	2
(3, 6)	3-2-6	3-2-6	2
(4, 2)	4-1-2	4-1-2	2
(4, 5)	4-1-5	4-1-4-5	3
(5, 1)	5-1	5-4-1	2
(5, 3)	5-6-2-3	5-6-2-3	3
(5, 6)	5-6	5-6	1
(6, 2)	6-2	6-2	1
(6, 4)	6-2-1-4	6-2-1-4	3

Table 4.4

Node pair	Virtual route	Physical route	Number of hops
(1, 2)	1-4-2	1-4-2	2
(1, 4)	1-4	1-4	1
(1, 6)	1-3-2-6	1-3-2-6	3
(2, 3)	2-3	2-3	1
(2, 5)	2-4-5	2-4-5	2
(3, 1)	3-1	3-1	1
(3, 4)	3-2-4	3-2-4	2
(3, 6)	3-2-6	3-2-6	2
(4, 2)	4-2	4-2	1
(4, 5)	4-5	4-5	1
(5, 1)	5-4-1	5-4-1	2
(5, 3)	5-6-2-3	5-6-2-3	3
(5, 6)	5-6	5-6	1
(6, 2)	6-2	6-2	1
(6, 4)	6-5-4	6-5-4	2

Simulated-Annealing-Based Heuristic

A heuristic based on a simulated annealing algorithm [25] for node mapping with a hypercube was proposed in [5]. The objective of this heuristic is also to minimize the total average propagation delay in the virtual topology. The heuristic can be briefly described as follows.

- The node mapping process starts with an initial random mapping. This initial mapping is then refined in an iterative manner by swapping adjacent nodes in the virtual topology.
- In each iteration, adjacent nodes are examined for swapping and a swapping of two adjacent nodes results in a new mapping. For example, suppose that node i is connected to node j. Meanwhile, node a and node b are adjacent to node i, while node c and node d are adjacent to node j. After node i is swapped with node j, node a and node b become adjacent to node j, while node c and node d become adjacent to node i, resulting in a new mapping. The average propagation delay is then computed over the new mapping.
- This mapping is accepted if it results in a smaller average propagation delay than the old mapping. Otherwise, it is accepted only with a probability that is determined by a control parameter. However, this probability decreases as the node-mapping process proceeds so as to simulate the cooling process associated with annealing.

- The lightpath routing and wavelength assignment subproblems are solved in a similar way to that described in the greedy-based heuristic.

4.4.4 Heuristics for Predetermined Topology Design

For predetermined virtual topology design, the virtual topology is supposed to be known a priori in terms of a set of lightpaths with their source and destination nodes. This means that the topology design subproblem has been solved. In many cases, it is assumed that the traffic pattern in the network has already been taken into account in the topology design subproblem and thus the traffic routing subproblem has also been solved. Therefore, only the lightpath routing and wavelength assignment subproblems need to be addressed. This virtual topology design problem was addressed in [11]. The routing and wavelength assignment subproblems are formulated as a multicommodity problem, in which a commodity is equivalent to a lightpath or a flow. The total number of flows on a fiber link gives the number of wavelengths needed on the link. The objective is to route each lightpath over the physical topology and assign a wavelength to each lightpath so that the maximum number of flows (i.e., wavelengths) on a fiber link can be minimized. For this purpose, randomized rounding and graph coloring can be used to solve the lightpath routing and wavelength assignment subproblems, respectively.

Randomized Rounding-Based Heuristic

The randomized rounding technique [11] is used to decide a physical route for each lightpath in the virtual topology. The basic principle is to relax integer constraints on the flows in the integer linear programming (ILP) problem and then find an approximate solution to the ILP problem by solving a linear programming (LP) problem. This heuristic algorithm consists of the following three phases.

- In the first phase, the integer constraints on the flows in the ILP formulation are relaxed to allow fractional flows. The relaxed problem becomes an LP problem and can thus be solved by some linear programming algorithm.
- In the second phase, a so-called path stripping process is carried out, in which a set of possible routes is created for each commodity (or lightpath). The set of possible routes for a commodity is initially set to be empty. A route for the commodity is then computed, which consists of a set of links that carry any part of the flow of that commodity. This route is then added to the set of possible routes along with a weight

equal to the minimum flow carried on a link of the route. To update, the flow on each link of the route is subtracted by the minimum flow, and all the links that carry no flow are removed. This process continues until there is no flow from the source node of the commodity.

- In the third phase, a route is chosen for each lightpath from the set of possible routes created in the second phase. This route is chosen randomly by using the weights assigned during the path stripping process.

Graph Coloring

The graph coloring technique is used to assign a wavelength to each of the lightpaths. This heuristic consists of the following two phases.

- In the first phase, a graph is constructed for the virtual topology or a set of lightpaths, which physical paths have been computed by the randomized rounding heuristic. Each lightpath corresponds to a node in the graph. If two lightpaths share a common physical link, the corresponding nodes are connected by an undirected edge in the graph.
- In the second phase, each node in the graph is colored in such a way that no two adjacent nodes have the same color. These colors correspond to the wavelengths used on the lightpaths in the virtual topology. Accordingly, the objective is to minimize the number of colors used for coloring the graph. The minimum number of colors required to color the nodes of a graph in such a manner is called the chromatic number of the graph.

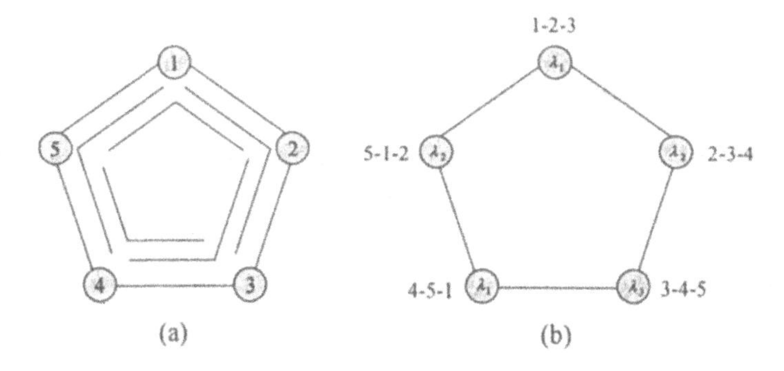

Figure 4.8 Illustration of graph coloring for wavelength assignment.

Example 4.5: Consider a physical topology with five nodes as shown in Figure 4.8(a). Assume that a virtual topology consists of five lightpaths, which physical routes have been computed by the randomized rounding heuristic, also shown in Figure 4.8(a). To solve the wavelength assignment subproblem, a graph is first constructed and graph coloring is performed following the steps described in the graph coloring heuristic. One solution to the wavelength assignment subproblem is shown in Figure 4.8(b).

The graph coloring problem is known to be NP complete. To address this problem, an efficient sequential graph-coloring algorithm called smallest-last coloring [16] can be used to obtain approximate solutions to the problem. The readers are referred to [16] for details.

4.4.5 Heuristics for Arbitrary Topology Design

For arbitrary virtual topology design, the topology design subproblem itself as well as some or all of the subsequent subproblems need to be addressed. In this context, many heuristic algorithms are proposed in the literature [7][9][22–24]. Here we only present three well-known heuristic algorithms.

Heuristic Logic Topology Design Algorithm

The heuristic logic topology design algorithm (HLDA) [6] is a simple heuristic algorithm for the logic topology design. The objective is to minimize the congestion in the network. This heuristic algorithm attempts to determine a set of lightpaths in terms of their source and destination nodes in the order of descending traffic demand in the belief that maximizing one-hop traffic may result in lower congestion in the network. Given the physical topology, the traffic demand matrix, and the logical degree of the network, HLDA can be briefly described as follows.

- Select a pair of source and destination nodes with the largest traffic demand in the traffic matrix.
- If there is a wavelength available on the shortest-delay path between the source and destination nodes, a transmitter available at the source node, and a receiver available at the destination node, a lightpath is established between the source and destination nodes. In this case, the corresponding traffic is subtracted by the next largest traffic in the traffic matrix.
- If there is no wavelength available on the shortest-delay path between the source and destination nodes, no transmitter available at the source node, or no receiver at the destination node, no lightpath is established

between the source and destination nodes. In this case, the corresponding traffic in the traffic matrix is set to zero.

- The above procedure is repeated until all node pairs with traffic demand are considered or the wavelength and degree constraints are broken.
- If there are nodes that still have transmitters and receivers available, establish as many lightpaths as possible at random as long as the wavelength and degree constraints are not broken.

To better understand the procedure of HLDA, let us examine the following example.

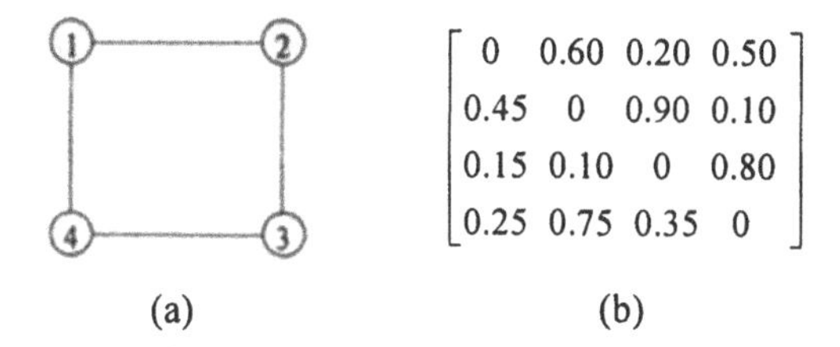

Figure 4.9 Illustration of HLDA (1): (a) physical topology; (b) original traffic matrix.

Example 4.6: Consider the physical topology shown in Figure 4.9(a). The original traffic matrix is given in Figure 4.9(b). Assume that there are two transmitters and two receivers at each node and there are two wavelengths on each fiber link. Figure 4.10 gives the traffic matrix and virtual topology in each iteration of HLDA. In the traffic matrices, the node pair with the largest traffic is indicated by bold italics. In iteration (1), the node pair with the largest traffic (i.e., 0.90) is selected, i.e., (2, 4). Thus, a lightpath is established between node 2 and node 4 on wavelength 1. The traffic associated with node pair (2, 4) is then subtracted by the next largest traffic (i.e., 0.80) and is set to 0.10. In iteration (2), the node pair with the largest traffic is selected, i.e., (3, 4). A lightpath is established between node 3 and node 4 also on wavelength 1. The traffic associated with node pair (3, 4) is then subtracted by the next largest traffic (i.e., 0.75) and is set to 0.05. In this manner, a lightpath is established in each iteration until all node pairs with traffic demand are considered or the wavelength and degree constraints are broken. Table 4.5 lists all the lightpaths established in this example. Note that in iteration (8), node pair (4, 1) is selected. However, there is no transmitter available at node 4. Thus, no lightpath is established between node 4 and node 1, and the traffic associated with node pair (4, 1) is set to zero. At the end of iteration

(9), there is no transmitter and receiver available at all the nodes. Therefore, no lightpath can be further established thereafter. The virtual topology finally obtained is shown in Figure 4.10(9).

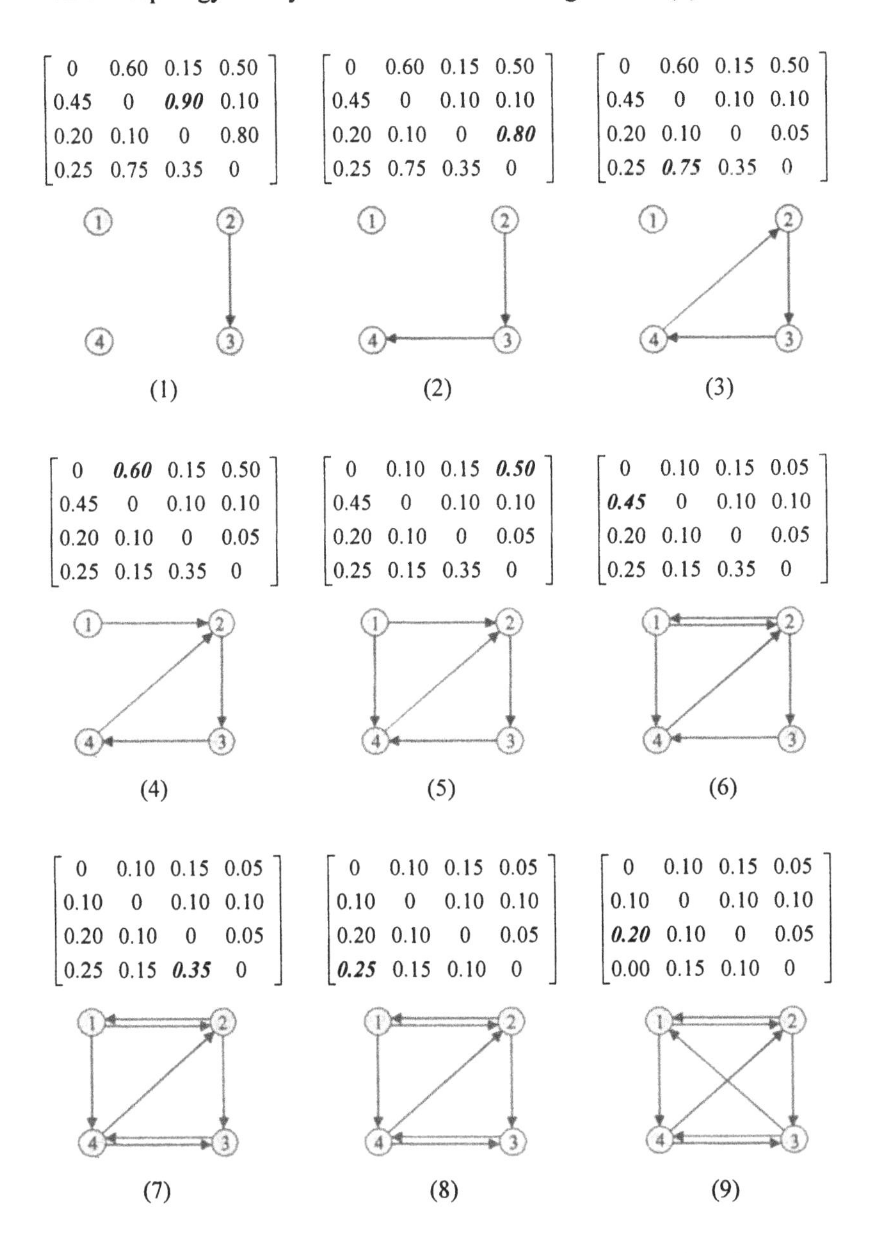

Figure 4.10 Illustration of HLDA (2): traffic matrix and virtual topology.

Table 4.5

Iteration	Lightpath	Route	Wavelength
(1)	2-3	2-3	1
(2)	3-4	3-4	1
(3)	4-2	4-3-2	1
(4)	1-2	1-2	1
(5)	1-4	1-4	1
(6)	2-1	2-1	1
(7)	4-3	4-3	2
(8)	None	None	None
(9)	3-1	3-4-1	2

Note that this heuristic may establish multiple lightpaths between a pair of nodes if the traffic between them, after subtracted by the next highest traffic, is still higher than the next highest traffic. The readers are referred to [6] for more details.

Minimum-Delay Logical Topology Design Algorithm

The minimum-delay logical topology design algorithm (MLDA) [6] is only applicable to the case in which the logical degree of the network is larger than the physical degree. The objective is to minimize the average propagation delay. MLDA can be briefly described as follows.

- Establish a pair of directed logical links or lightpaths in opposite directions for each physical link.
- Employ HLDA to add other logic links or lightpaths taking into account the traffic demand between each node pair.

This ensures that there is a shortest path for each pair of nodes and thus minimizes the average propagation delay between each pair of nodes. However, it may not be able to minimize the congestion in the network. The readers are referred to [6] for more details. The procedure of MLDA is further illustrated in the following example.

Example 4.7: Consider again the physical topology in Example 4.4, as shown in Figure 4.11(a). The original traffic matrix is shown in Figure 4.11(b). Assume that there are three transmitters and three receivers at each node and there are two wavelengths (i.e., wavelength 1 and wavelength 2) on each fiber link. Figure 4.12 gives the traffic matrix and virtual topology in each iteration of MLDA. In iteration (1), a pair of directed logical links is established in opposite directions for each physical link. In iterations (2), (3), and (4), a lightpath is added based on HLDA. At the end of iteration (4), no transmitter and receiver are available at all nodes. Therefore, no lightpath can be further added thereafter. The virtual topology obtained at the end of iteration (4) is the final one. Table 4.6 lists all the lightpaths established in this example.

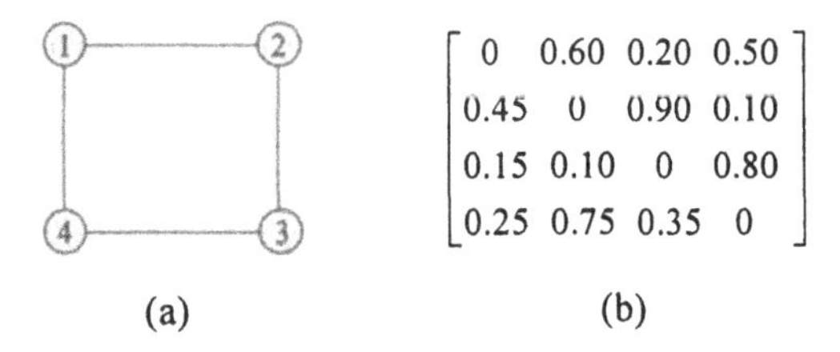

$$\begin{bmatrix} 0 & 0.60 & 0.20 & 0.50 \\ 0.45 & 0 & 0.90 & 0.10 \\ 0.15 & 0.10 & 0 & 0.80 \\ 0.25 & 0.75 & 0.35 & 0 \end{bmatrix}$$

(a) (b)

Figure 4.11 Illustration of MLDA (1): (a) physical topology; (b) original traffic matrix.

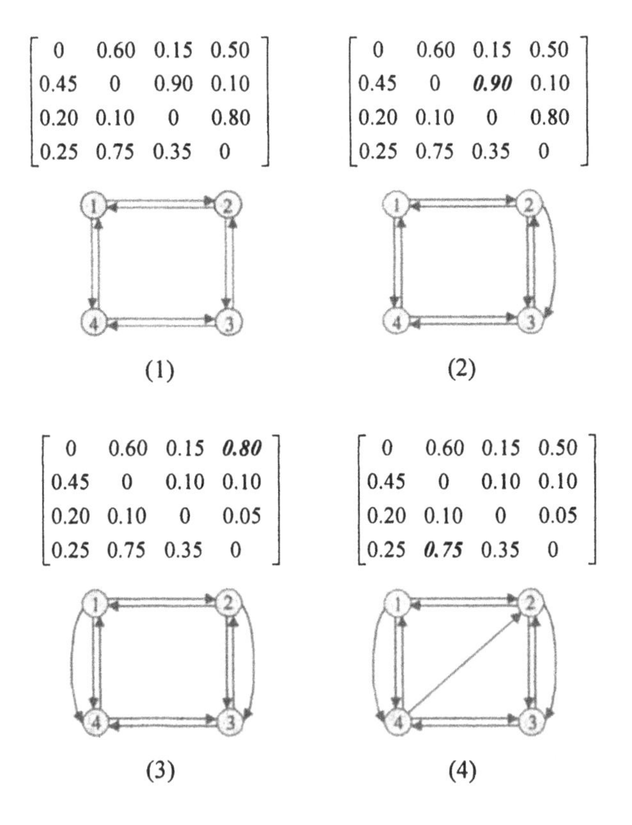

(1)

$$\begin{bmatrix} 0 & 0.60 & 0.15 & 0.50 \\ 0.45 & 0 & 0.90 & 0.10 \\ 0.20 & 0.10 & 0 & 0.80 \\ 0.25 & 0.75 & 0.35 & 0 \end{bmatrix}$$

(2)

$$\begin{bmatrix} 0 & 0.60 & 0.15 & 0.50 \\ 0.45 & 0 & \boldsymbol{0.90} & 0.10 \\ 0.20 & 0.10 & 0 & 0.80 \\ 0.25 & 0.75 & 0.35 & 0 \end{bmatrix}$$

(3)

$$\begin{bmatrix} 0 & 0.60 & 0.15 & \boldsymbol{0.80} \\ 0.45 & 0 & 0.10 & 0.10 \\ 0.20 & 0.10 & 0 & 0.05 \\ 0.25 & 0.75 & 0.35 & 0 \end{bmatrix}$$

(4)

$$\begin{bmatrix} 0 & 0.60 & 0.15 & 0.50 \\ 0.45 & 0 & 0.10 & 0.10 \\ 0.20 & 0.10 & 0 & 0.05 \\ 0.25 & \boldsymbol{0.75} & 0.35 & 0 \end{bmatrix}$$

Figure 4.12 Illustration of MLDA (2): traffic matrix and virtual topology.

Table 4.6

Iteration	Lightpath	Route	Wavelength
(1)	1-2	1-2	1
(1)	2-1	2-1	1
(1)	2-3	2-3	1
(1)	3-2	3-2	1
(1)	3-4	3-4	1
(1)	4-3	4-3	1
(1)	1-4	1-4	1
(1)	4-1	4-1	1
(2)	2-3	2-3	2
(3)	1-4	1-4	2
(4)	4-2	4-3-2	2

Traffic-Independent Logical Topology Design Algorithm

The traffic-independent logical topology design algorithm (TILDA) [6] addresses the problem without taking into account the traffic demand but attempts to establish lightpaths with as few physical links as possible. The objective is to minimize the number of wavelengths needed. TILDA can be briefly described as follows.

- Establish a logical link between all one-hop nodes in the physical topology.
- Establish a logic link between all two-hop nodes under the conditions that there are no lightpaths between them and the degree constraints are not broken.
- Establish a logic link between all three-hop nodes under the conditions that there are no lightpaths between them and the degree constraints are not broken.
- Continue in the same manner until the degree constraints are broken.

This heuristic algorithm can be a good choice if the traffic demand is not known or known to be uniform. The readers are referred to [6] for more details.

Example 4.8: Consider again the physical topology shown in Figure 4.2(a). Assume that there are three transmitters and three receivers at each node and there are two wavelengths (e.g., wavelength 1 and wavelength 2) on each fiber link. Figure 4.13 shows the virtual topology obtained in each iteration of TILDA. In iteration (1), a logical link is established between all one-hop nodes in the physical topology, as shown in Figure 4.13(a). In iteration (2), a logical link is established between all two-hop nodes in the physical topology, as shown in Figure 4.13(b). At the end of iteration (2), no transmitter and receiver are available at all the nodes. Therefore, no lightpath can be further established thereafter. Table 4.7 lists all the lightpaths established in this example.

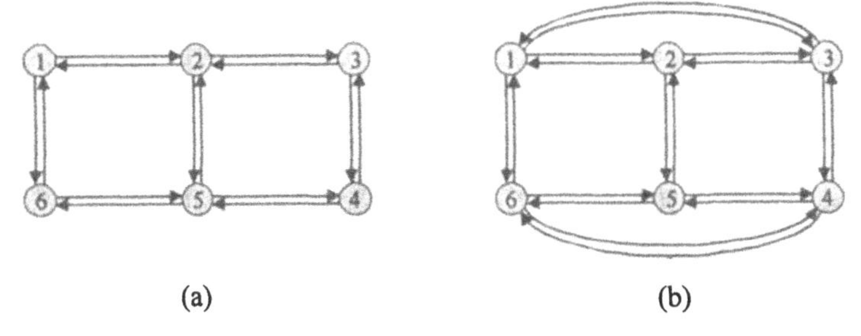

Figure 4.13 Illustration of TILDA: (a) iteration (1); (b) iteration (2).

Table 4.7

Iteration	Lightpath	Route	Wavelength
(1)	1-2	1-2	1
(1)	2-1	2-1	1
(1)	2-3	2-3	1
(1)	3-2	3-2	1
(1)	3-4	3-4	1
(1)	4-3	4-3	1
(1)	4-5	4-5	1
(1)	5-4	5-4	1
(1)	5-6	5-6	1
(1)	6-5	6-5	1
(1)	6-1	6-1	1
(1)	1-6	1-6	1
(2)	1-3	1-2-3	2
(2)	3-1	3-2-1	2
(2)	4-6	4-5-6	2
(2)	6-4	6-5-4	2

4.5 Virtual Topology Reconfiguration

A virtual topology consists of a set of lightpaths established over the physical topology, which is designed based on the traffic demand and physical topology of a network. However, the traffic demand and physical topology of the network cannot be constant without any change over time. In fact, there are various factors that may result in a change in traffic demand and physical topology. For this reason, a virtual topology may need to be redesigned and reconfigured in response to the changing traffic demand and physical topology in the network. Reconfiguration is a process by which a network is redesigned and reconfigured from an old virtual topology to a new virtual topology. The purpose of reconfiguration is to maintain the optimal network performance under various traffic and topology changes. The highly reconfigurable characteristic of wavelength-routed WDM networks has provided a good flexibility in implementing virtual topology reconfiguration in response to changing traffic demand and physical topology.

Reconfiguration problems

In general, there are two basic types of problems in virtual topology reconfiguration.

- Type 1 problem: The physical topology as well as the old virtual topology and the new virtual topology that the network must be reconfigured to are already known. The objective is to minimize the cost in the reconfiguration. The reconfiguration cost can be measured in terms of the number of OXCs that need to be reconfigured or the number of lightpaths that need to be changed during the reconfiguration. Such cost metrics can reflect the amount of time taken to make the changes and the service disruption incurred during the reconfiguration. Other metrics can also be used as long as they can reasonably reflect the amount of time taken or service disruption incurred for making the reconfiguration. Several heuristic algorithms have been proposed for solving this type of reconfiguration problem [17–19].
- Type 2 problem: Only the old virtual topology as well as the new traffic demand or new physical topology is already known. The objective is to design a new virtual topology based on an objective function used in the virtual topology design as well as an additional objective function to minimize the cost of the reconfiguration. The reconfiguration cost can be measured in terms of some metrics such as the number of lightpaths added or deleted, or the number of disrupted lightpaths, to reflect the amount of time taken or the service disruption incurred for making the

reconfiguration. To solve this type of reconfiguration problem, a number of heuristic algorithms have been proposed in the literature [7][20–21].

Reconfiguration methods

Like the virtual topology design, there are two basic methods for solving the reconfiguration problem: linear programming and heuristics.

- Linear programming: The virtual topology reconfiguration problem can be also formulated as an MILP problem. The formulation is similar to that of the virtual topology design problem, usually with an additional objective function to minimize the reconfiguration cost. An optimal solution to MILP gives an exact solution to the reconfiguration problem. However, because MILP is computationally intractable, this method is only applicable to small networks. In fact, even for small networks, a huge amount of computational time is required to solve MILP. For this reason, linear programming is usually unsuitable for solving the reconfiguration problem.
- Heuristics: Heuristics can be used to obtain solutions to the reconfiguration problem. Because heuristics are computationally tractable and can provide good approximate solutions, this method has been used widely for solving the virtual topology reconfiguration problem.

On the other hand, the reconfiguration can be carried out either off-line or on-line.

- Off-line reconfiguration: The network is turned off, and thus all lightpaths are disrupted. Obviously, this can result in a large amount of traffic disruptions in the network, which is not practical.
- On-line reconfiguration: The network is not turned off. Only the lightpaths that need to be changed are disrupted and the traffic disruptions can thus be largely reduced. Because the traffic disruptions depend on the number of lightpath changes, the reconfiguration problem must consider an objective function that minimizes the total number of lightpath changes in the reconfiguration.

In the next sections, we will discuss the reconfiguration problem for traffic changes and topology changes, respectively, in more detail.

4.5.1 Reconfiguration for Traffic Changes

A virtual topology is designed based on the traffic demand of a network. The traffic demand comes from the higher layers of the optical layer. Although such traffic demand can keep relatively stable in the short term in large transport networks, it is subject to change in the long term. As a result, the virtual topology that is optimal for the old traffic demand may be no longer optimal for the new traffic demand. This may greatly affect the network performance and thus degrade the network service. For this reason, there is a need for reconfiguring the old virtual topology to a new virtual topology to address the changing traffic demand.

In general, there are three basic strategies to deal with traffic changes in the network, which are described as follows.

- Strategy 1: Do not change the virtual topology, but use the same virtual topology for any traffic demand. With this strategy, no lightpath and traffic disruptions would be incurred. However, it is unable to deal with traffic changes and thus unable to maintain optimal network performance under traffic changes. Therefore, this strategy is not reasonable and is not preferred.
- Strategy 2: Design a new virtual topology based on the new traffic demand without taking the old virtual topology into account. With this strategy, although the new virtual topology can provide the optimal performance to meet the new traffic demand, it might need to change a large number of lightpaths in order to make the reconfiguration. As a result, a large amount of lightpath disruption might be incurred, which would result in a large amount of service disruption.
- Strategy 3: Design a new virtual topology based on the new traffic demand as well as the old virtual topology with an optimization objective and an additional objective to minimize the number of lightpath changes for the reconfiguration. This strategy can provide a trade-off between network performance and network service.

As mentioned earlier, both linear programming and heuristics can be used to solve the reconfiguration problem for traffic changes. An example of linear programming formulations for solving such a problem can be found in [7]. An example of heuristics for solving such a problem can be found in [20].

4.5.2 Reconfiguration for Topology Changes

In general, a physical topology change may be caused by a network failure or a topology update. However, network providers are usually more concerned with topology changes as a result of network failures because such changes happen more often. A network failure can be a fiber cut or a node fault. In the event of a network failure, many lightpaths might be disrupted, which would cause a large amount of traffic disruption and data loss and thus largely degrade the network performance.

To deal with a network failure, there are usually two basic strategies: rerouting and reconfiguration.

- Rerouting: The rerouting strategy is to find an alternate route for each of the lightpaths that are disrupted without affecting other lightpaths. However, this strategy may not be able to find an optimal route or even a route for each disrupted lightpath in the event of a network failure. In general, rerouting can be performed on a link basis or on a path basis, as shown in Figure 4.14(a) and Figure 4.14(b), respectively, which is not the focus of this chapter. A more detailed discussion of rerouting for a network failure will be given in Chapter 6.
- Reconfiguration: The reconfiguration strategy is to reconfigure a new virtual topology based on the new physical topology. This strategy can provide better routes for disrupted lightpaths and can thus handle more disrupted lightpaths. However, it may affect those lightpaths that are not disrupted by the failure during the reconfiguration.

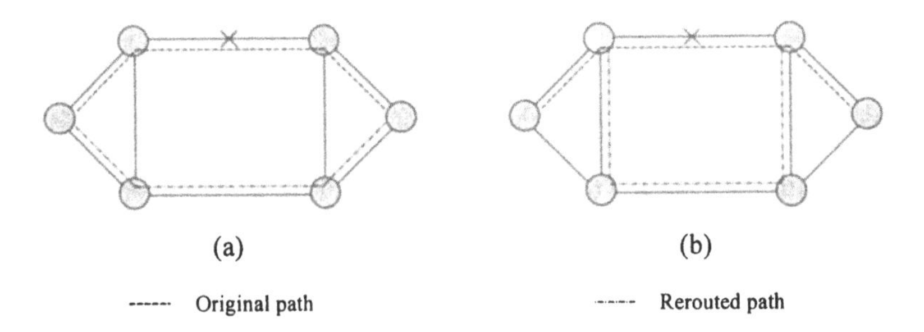

Figure 4.14 Rerouting policies: (a) on a path basis; (b) on a link basis.

The basic methods used to solve the reconfiguration problem for network failures are similar to those used for traffic changes as discussed in Section 4.5.1.

4.6 Summary

Virtual topology design is an important problem in wavelength-routed WDM networks. A virtual topology combines the advantages of optical transmission and electronic processing, and can therefore enhance network characteristics and improve network performance. In this chapter, the virtual topology design problem was described and the fundamentals of virtual topology design were introduced. The virtual topology design problem can be formulated as an MILP problem, and an exact formulation of the virtual topology design problem was presented with an objective function to minimize the congestion of the network. The formulated problem is known to be NP hard and is thus computationally intractable. To make the problem tractable, the whole problem can be decomposed into four subproblems and heuristics can be used to obtain approximate solutions to the whole problem or its subproblems. To evaluate an approximate solution, lower bounds on network congestion and the number of wavelengths needed in the virtual topology design were discussed. Heuristics design with regular topology, predetermined topology, and arbitrary topology was also discussed and several well-known heuristics for solving different subproblems were presented.

This chapter also discussed the virtual topology reconfiguration problem, which is similar to the virtual topology design problem. The need for reconfiguration was explained, and the reconfiguration problem was described. Moreover, basic reconfiguration strategies and methods for solving the reconfiguration problem in general and for handling traffic changes and topology changes in particular were discussed.

Problems

4.1 What is the virtual topology design problem? Why is this problem important in wavelength-routed WDM networks?

4.2 What are the four subproblems that a virtual topology design problem can be decomposed into? Describe the objectives of these subproblems.

4.3 Consider the physical topology in Figure 4.3. Suppose that there are two transmitters and two receivers at each node and two wavelengths on each fiber link. Give another example to explain the four subproblems of a virtual topology design problem.

4.4 Explain what kinds of network constraints should be considered in the formulation of a virtual topology design problem.

4.5 Why is it necessary to know the theoretical lower bounds on the optimal solution to the virtual topology design problem? Describe the lower bounds on the network congestion and the number of wavelengths.

4.6 Consider Example 4.6. Assume that the highest traffic demand in the traffic matrix is 1.60 rather than 0.95. What is the resulting virtual topology? Illustrate the trace of traffic matrix and virtual topology in each iteration.

4.7 Consider Example 4.8. Assume that there are four transmitters and four receivers at each node and there are three wavelengths on each fiber link. Design the virtual topology with TILDA.

4.8 What is the virtual topology reconfiguration problem? Why is it necessary and important?

4.9 What are the two basic types of problems in the virtual topology reconfiguration?

4.10 What are the basic strategies to deal with the traffic changes? Explain their advantages and disadvantages.

4.11 What are the basic strategies to deal with the network failures? Explain their advantages and disadvantages.

4.12 What are the two different methods for solving the virtual topology design and reconfiguration problems? Explain their advantages and disadvantages.

References

[1] Rajiv Ramaswami and Kumar N. Sivarajan, *Optical Networks–A Practical Perspective*, Second Edition, Morgan Kaufmann Publishers, San Francisco, 2002.

[2] R. Dutta and G. N. Rouskas, "A survey of virtual topology design algorithms for wavelength routed optical networks," *SPIE Optical Networks Magazine*, vol. 1, no. 1, Jan. 2000, pp. 73–89.

[3] C. Siva Ram Murthy and Mohan Gurusamy, *WDM Optical Networks: Concepts, Design, and Algorithms*, Prentice Hall PTR, Upper Saddle River, New Jersey, 2002.

[4] B. Mukherjee, S. Ramamurthy, D. Banerjee, and A. Mukherjee, "Some principles for designing a wide-area optical network," *Proceedings of IEEE INFOCOM'94*, pp. 110–119.

[5] B. Mukerjee, D. Banerjee, S. Ramamurthy, and A. Mukherjee, "Some principles for designing a wide-area WDM optical network," *IEEE/ACM Transactions on Networking*, vol. 4, no. 5, Oct. 1996, pp. 648–696.

[6] R. Ramaswami and K. N. Sivarajan, "Design of logical topologies for wavelength-routed optical networks," *IEEE Journal on Selected Areas in Communications*, vol. 14, no. 5, Jun. 1996, pp. 840–851.

[7] D. Banerjee and B. Mukherjee, "Wavelength-routed optical networks, linear formulation, resource budgeting tradeoffs, and a reconfiguration study,"

IEEE/ACM Transactions on Networking, vol. 8, no. 5, Oct. 2000, pp. 598–607.

[8] R. M. Krishnaswamy and K. N. Sivarajan, "Design of logical topologies: a linear formulation for wavelength-routed optical networks with no wavelength changers," *IEEE/ACM Transactions on Networking*, vol. 9, no. 2, Apr. 2001, pp. 186–198.

[9] R. M. Krishnaswamy and K. N. Sivarajan, "Design of logical topologies: a linear formulation for wavelength routed optical networks with no wavelength changers," *Proceedings of IEEE INFOCOM'98*, pp. 919–927.

[10] M. A. Marsan, A. Bianco, E. Leonardi, and F. Neri, "Topologies for wavelength-routing all-optical networks," *IEEE/ACM Transactions on Networking*, vol. 1, no. 5, Oct. 1993, pp. 534–546.

[11] D. Banerjee and B. Mukherjee, "A practical approach for routing and wavelength assignment in large wavelength-routed optical networks," *IEEE Journal on Selected Areas in Communications*, vol.14, no. 5, Jun. 1996, pp. 903–908.

[12] I. Chlamtac, A. Ganz, and G. Karmi, "Lightpath communications: An approach to high bandwidth optical WANs," *IEEE Transactions on Communications*, vol. 40, no. 7, Jul. 1992, pp. 1171–1182.

[13] J.-F. P. Labourdette and A. S. Acampora, "Logically rearrangeable multihop lightwave networks," *IEEE Transactions on Communications*, vol.39, no. 8, Aug. 1991, pp. 1223–1230.

[14] D. Bienstock and O. Gunluk, "Computational experience with a difficult mixed-integer multicommodity flow problem," *Mathematical Programming*, vol. 68, 1995, pp. 213–237.

[15] I. Chlamtac, A. Ganz, and G. Karmi, "Lightnets: Topologies for high-speed optical networks," *Journal of Lightwave Technology*, vol. 11, no. 5/6, May/Jun. 1993, pp. 951–961.

[16] D. W. Matula, G. Marble, and J. D. Isaacson, "Graph coloring algorithms," *Graph Theory and Computing*, R. C. Read (Ed.), Academic Press, New York and London, 1972, pp. 109–122.

[17] D. Bienstock and O. Gunluk, "A degree sequence problem related to network design," *Networks*, vol. 24, no. 4, Jul. 1994, pp. 195–205.

[18] J.-F. P. Labourdette, G. W. Hart, and A. S. Acampora, "Branch-exchange sequences of reconfiguration of lightwave networks," *IEEE Transactions on Communications*, vol. 42, no. 10, Oct. 1994, pp. 2822–2832.

[19] G. N. Rouskas and M. H. Ammar, "Dynamic reconfiguration in multihop WDM networks," *Journal of High Speed Networks*, vol. 4, no. 3, 1995, pp. 221–238.

[20] N. Sreenath, C. Siva Ram Murthy, B. H. Gurucharan, and G. Mohan, "A two-stage approach for virtual topology reconfiguration of WDM optical networks," *SPIE Optical Networks Magazine*, vol. 2, no. 3, May/Jun. 2001, pp. 58–71.

[21] Jun Zheng, Bin Zhou, and Hussein T. Mouftah, "Dynamic reconfiguration based on balanced alternate routing (BARA) for all-optical wavelength-routed WDM networks," *Proc. IEEE GLOBECOM'02*, vol. 3, 2002, pp. 2706–2710.

[22] S. Banerjee, J. Yoo, and C. Chen, "Design of wavelength-routed optical networks for packet switched traffic," *Journal of Lightwave Technology*, vol. 15, no. 9, Sep. 1997, pp. 1636–1646.

[23] Z. Zhang and A. Acampora, "A heuristic wavelength assignment algorithm for multihop WDM networks with wavelength routing and wavelength re-use," *IEEE/ACM Transactions on Networking*, vol. 3, no. 3, Jun. 1995, pp. 281–288.

[24] Anwar Haaque, Yash P. Aneja, Subir Bandyopadhyay, Arunita Jaekel, and Abhijit Sengupta, "Some studies on the logical topology design of large multi-hop optical networks," *SPIE Optical Networks Magazine*, vol. 3, no. 4, Jul./Aug. 2002, pp. 96–105.

[25] A. Aarts and J. Korst, *Simulated Annealing and Boltzmann Machines*, John Wiley and Sons, New York, 1989.

Chapter 5

Distributed Lightpath Establishment

5.1 Introduction

In Chapter 3, we discussed the routing and wavelength assignment (RWA) problem for lightpath establishment under centralized control. The advantage of centralized control is the simplicity in control. Because the centralized controller of a network maintains global network state information for controlling the whole network, there is no need to have a high degree of coordination among different network nodes. However, centralized control is not reliable. A failure with the centralized controller can result in the breakdown of the whole network and thus disrupt all network services. Moreover, the centralized controller must maintain a large database to control all nodes, links, and connections in the network, which is not scalable. For these reasons, centralized control is considered unsuitable for large networks with dynamic traffic. To overcome the shortcomings of centralized control, distributed control is highly preferred because it can significantly improve network scalability and reliability. However, distributed control increases the difficulty and complexity in control, which presents more challenges in the network design.

This chapter focuses on distributed lightpath establishment in wavelength-routed WDM networks. The main problems with distributed lightpath establishment are first discussed, and the basic routing, wavelength assignment, and wavelength reservation paradigms are then introduced.

Based on these basic paradigms, several typical distributed control protocols for lightpath establishment are presented. We assume no wavelength conversion at each network node unless otherwise stated.

5.2 Problems in Distributed Lightpath Establishment

The architecture of a wavelength-routed WDM network has been shown in Figure 1.6. Under distributed control, the logic architecture of the network consists of a data network and a control network, as shown in Figure 5.1. The data network (in solid lines) consists of optical switches and data channels, which are used for transferring data, and operates in circuit switching. The data channels are implemented by a certain number of optical channels or wavelengths on each fiber link. The control network (in dotted lines) consists of electronic controllers and control channels, which are used for exchanging network state information and signaling-control information, and operates in packet switching. The control channels can be implemented either by one or more dedicated wavelengths over each fiber link or through a dedicated IP network.

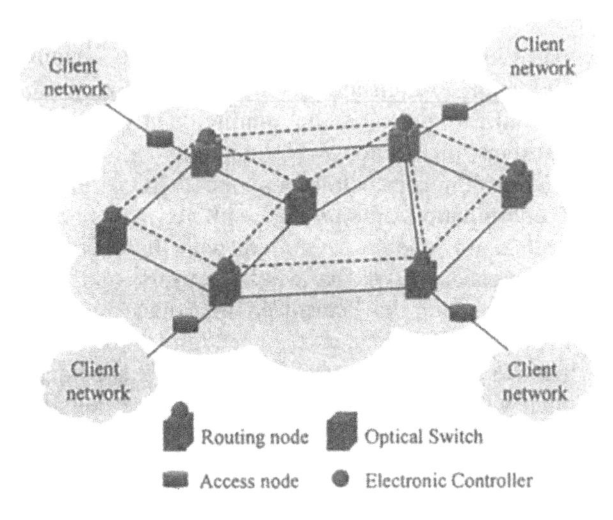

Figure 5.1 Logic architecture of a wavelength-routed WDM network.

Under distributed control, all connection requests are processed at different network nodes concurrently and each node makes its decisions independently based on the network state information it maintains, which can be either local or global. To dynamically establish a lightpath for each connection request, the network must first decide on a physical route for the connection and select a suitable wavelength on each link along the decided route and then reserve the selected wavelength and configure the OXC at each intermediate node along the decided route. In the presence of wavelength converters, a suitable wavelength can be any available wavelength on a fiber link. Otherwise, all the links along a decided route must use the same wavelength. If there is no wavelength available on any of the links or there is no common wavelength available on all the links in the absence of wavelength converters, a connection request will be blocked. Accordingly, distributed lightpath establishment involves three main problems:

- Routing
- Wavelength assignment
- Wavelength reservation

5.2.1 Routing

The objective of routing is to decide a physical route for each connection request in a manner that can improve the network performance in terms of request blocking probability and connection setup time. This can usually be achieved by performing a routing algorithm at a network node. The request blocking probability, or simply blocking probability, is defined as the ratio of the number of successful connection requests to the total number of connection requests to the network. The connection setup time is the time taken from the instant that a connection request arrives to the network to the instant that a lightpath is successfully established for the connection request. Because the number of wavelengths available on each fiber link is limited and there is the wavelength-continuity constraint in the absence of wavelength converters, a routing algorithm may have a great impact on the blocking probability and a good routing algorithm must be able to reduce the blocking probability. The computational complexity of a routing algorithm is another concern. Under distributed control, a routing algorithm is usually performed on-line at a network node. A higher computational complexity would result in a longer computational time and thus increase the connection setup time. For this reason, a routing algorithm must avoid high computational complexity.

Under distributed control, routing is based on the network state information maintained at each network node, which can be classified into three basic types:

- Local information
- Global information
- Neighborhood information

Routing based on local information

With local information, each node only knows the wavelength usage on its outgoing links. A routing decision is made without full knowledge of the wavelength usage on other links. As a result, a connection request is very likely to be blocked at each intermediate node of a route because of the unavailability of wavelength resources. This would result in a higher blocking probability in the network. However, routing based on local information is simpler to implement. There is no need to update the state information maintained at each network node on a global basis.

Routing based on global information

An effective way to overcome the shortcomings of using local information is to maintain global information at each network node. With global information, each node knows the network topology and wavelength usage on each link in the network. Accordingly, it can make more intelligent routing decisions, which can thus reduce blocking probability in the network. However, because the network state changes constantly under dynamic traffic, the network must employ a state advertisement protocol to timely update the global network state information maintained at each node in order to make correct decisions. This can be implemented either periodically or on a change [1], which would cause significant control overhead in the network. Such control overhead is also a big concern in distributed lightpath control. It should be pointed out that because of the characteristics of dynamic traffic as well as the propagation delay on each link, a network node may still make an incorrect decision based on updated information.

Routing based on neighborhood information

To achieve a trade-off between network performance and control overhead, routing based on neighborhood information [2] may provide a practical solution, in which the source node only maintains the wavelength usage information on the first *k* links of all possible routes to the destination node. The neighborhood information can be collected on demand or updated periodically. It was shown in [2] that routing based on neighborhood information can achieve good network performance in terms of the blocking probability compared to using local information. Meanwhile, it requires less control overhead for updating neighborhood information than for updating global information.

To achieve good network performance, a number of routing paradigms have been proposed in the literature, such as explicit routing, hop-by-hop routing, flooding-based routing, source routing, and destination routing. These routing paradigms will be described in more detail in Section 5.3.

5.2.2 Wavelength Assignment

The objective of wavelength assignment is to select a suitable wavelength among multiple available wavelengths in a manner that can increase wavelength utilization and reduce the blocking probability in the network. This is particularly important in the absence of wavelength converters. For this reason, a variety of wavelength assignment algorithms have already been proposed in the literature, such as Random, First-Fit, Most-Used, and Least-Used. As these algorithms have already been described in Section 3.4, we will not go into unnecessary details in this chapter.

5.2.3 Wavelength Reservation

The objective of wavelength reservation is to minimize the possibility of a reservation failure and to minimize wavelength reservation time. Because the network state changes constantly under dynamic traffic, a wavelength that is available at a given time may be no longer available at a later time. For this reason, a reservation failure is very likely to occur at an intermediate node along a decided route during wavelength reservation. On the other hand, because it takes a two-way delay to make wavelength reservation between a pair of source and destination nodes, the wavelength reservation time is also a concern. These factors should be taken into account in the design of a wavelength reservation protocol in order to minimize the possibility of a reservation failure and to minimize wavelength reservation time.

It should be pointed out that in many situations routing, wavelength assignment, and wavelength reservation are closely related to each other and are considered together, especially the routing and wavelength assignment problems. For example, in adaptive routing based on global information, routing and wavelength assignment are often performed together whereas wavelength reservation is relatively independent.

Distributed lightpath control has received much attention in recent years and has been widely studied in the literature. In the subsequent sections, we will introduce various basic routing and wavelength reservation paradigms that have already been proposed in the literature for distributed lightpath establishment.

5.3 Routing

Typically, there are three basic routing paradigms for distributed lightpath establishment:

- Explicit routing
- Hop-by-hop routing
- Flooding-based routing

5.3.1 Explicit Routing

In explicit routing, the entire route of a connection is decided by a single network node. This node can be either a source node or a destination node, which corresponds to source routing and destination routing, respectively. In source routing, the entire route of a connection is decided by the source node, whereas in destination routing, the entire route of a connection is decided by the destination node. A routing decision can be based on either local or global network state information. To achieve better performance, however, global information is highly preferred. Examples of source routing and destination routing based on global information can be found in [1] and [3], respectively. Explicit routing can further be classified into three basic routing paradigms:

- Fixed routing
- Fixed-alternate routing
- Adaptive routing

Fixed routing

In fixed routing, there is only a single fixed route between each pair of source and destination nodes in the network. This fixed route is precomputed off-line, and any connection between a pair of nodes must use the same fixed route. A typical example of fixed routing is the fixed shortest-path routing algorithm, which was described in Section 3.4.3.

Fixed-alternate routing

In fixed-alternate routing [4–5], there are a set of alternate routes between each pair of source and destination nodes. These alternate routes are precomputed off-line and are orderly stored in a routing table maintained at each node. The actual route of a connection can only be chosen from these alternate routes. A typical example of fixed-alternate routing is the K-shortest-path routing algorithm described in Section 3.4.3.

Adaptive routing

In adaptive routing [6–8], there is no restriction on routing. Any possible route between a pair of source and destination nodes can be chosen as the actual route of a connection. The choice of routes is based on the current network state information and a path selection policy. A typical example of adaptive routing is the least-cost path routing algorithm described in Section 3.4.3.

In general, fixed routing is the simplest paradigm to implement but can result in poor network performance in terms of blocking probability. Adaptive routing can achieve good network performance but is computationally complex to implement. Fixed-alternate routing provides a trade-off between computational complexity and network performance. These routing paradigms are similar to those described for centralized control in Section 3.4. The main difference is that under centralized control a routing algorithm is performed by a centralized controller in the network, whereas under distributed control a routing algorithm is performed by each network node.

5.3.2 Hop-by-Hop Routing

In hop-by-hop routing, each node independently decides the next hop of a route. A routing decision is made dynamically one hop at a time based on either local or global network state information.

Hop-by-hop routing based on global information

Hop-by-hop routing can be based on global network state information. With global information, a node can make a more intelligent decision to decide the next hop and can thus achieve better network performance. However, to maintain global information at each node, the network must employ a state advertisement protocol to update the global information in a timely manner, which would increase the complexity in implementation and introduce additional control overhead in the network.

An example of hop-by-hop routing based on global information is the distance-vector approach or distributed-routing approach proposed in [9]. In this approach, each node maintains a routing table for each wavelength, which indicates the next hop and the cost associated with the shortest path to each destination node on that wavelength. The cost can be measured in terms of hop counts or link distances. A connection request is routed one hop at a time. Each node independently decides the next hop, and the wavelength that results in the lowest cost is selected. The routing table is maintained with a distributed Bellman–Ford algorithm and is updated whenever a connection is set up or taken down. It has been shown that the distributed-routing approach yields lower blocking probability than the link-state approach in [1].

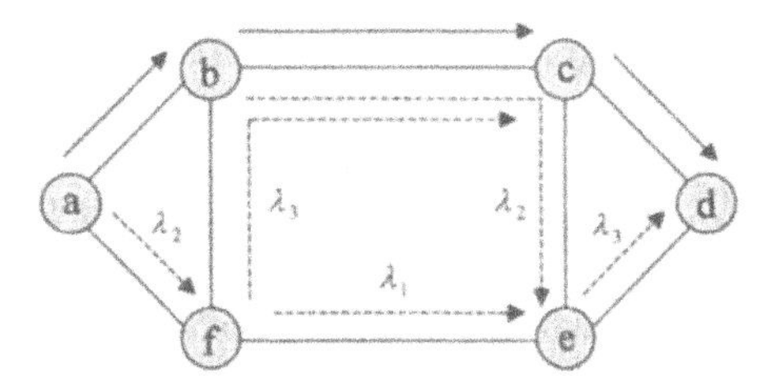

Figure 5.2 Illustration of distributed-routing approach.

Example 5.1: Consider a network consisting of six nodes and eight links with three wavelengths on each link, as shown in Figure 5.2. Suppose that there are five connections already established in the network. The routing table at each node maintains the routing information on the next hop and the cost associated with the shortest path to node *d* for each wavelength, which is shown in Table 5.1. Note that the cost in the routing tables is measured in terms of hop counts. Now assume that there is a connection request from node *a* to node *d*. To establish a connection for the request, node *a* will first decide node *b* as the next hop and select wavelength 1 because the cost associated with their shortest paths to node *d* for wavelength 1 is the lowest. Then node *a* forwards the connection request to node *b*. At node *b*, because the cost associated with the shortest path to node *d* for wavelength 1 is smaller than that for both wavelength 2 and wavelength 3, node *c* is decided as the next hop and wavelength 1 is selected, and the connection request is forwarded to node *c*. Similarly, node *c* will decide node *d* as the next hop and select wavelength 1 as all the wavelengths are available on the link directly to node *d*. As a result, the connection can be established on route *a-b-c-d* using wavelength 1.

Table 5.1 Routing information for node *d* at each node

Wavelength	Destination	Next hop	Cost
1	*d*	*b*	3
2	*d*	*b*	4
3	*d*	*f*	4

(a) Node *a*

Wavelength	Destination	Next hop	Cost
1	*d*	*c*	2
2	*d*	*f*	3
3	*d*	*a*	5

(b) Node *b*

Wavelength	Destination	Next hop	Cost
1	*d*	*d*	1
2	*d*	*d*	1
3	*d*	*d*	1

(c) Node *c*

Wavelength	Destination	Next hop	Cost
1	*d*	*d*	1
2	*d*	*d*	1
3	*d*	*d*	2

(d) Node *e*

Wavelength	Destination	Next hop	Cost
1	*d*	*b*	3
2	*d*	*e*	2
3	*d*	*e*	3

(e) Node *f*

Hop-by-hop routing based on local information

Hop-by-hop routing can also be based on local network state information. With local information, a node is unable to make a more intelligent decision to decide the next hop and thus cannot achieve the best performance. However, because there is no need to update global network state information, it is simpler to implement and has no additional control overhead introduced.

An example of hop-by-hop routing based on local information is the alternate-link routing approach or deflection routing approach proposed in [6]. In this approach, each node maintains a routing table, which indicates one or more alternate outgoing links to each destination node. These alternate links are precomputed and may be orderly stored in the routing table. Each node only needs to maintain local information about the wavelength availability on its outgoing links. A connection request is routed one hop at a time. Each node independently chooses one link from the alternate links based on some selection policy, such as the shortest path first or the least-congested link first, and then forwards the connection request to the next hop.

Under the shortest-path policy, each node first chooses the outgoing link that leads to the shortest path to the destination node. If there is no wavelength available on the link, the node will choose an alternate outgoing link that leads to the next shortest path to the destination node. If the wavelength is available, the node will forward the request to the next hop. If there is no wavelength available on all alternate links, the request will be blocked and dropped. The request proceeds in this manner until it reaches the destination node or it is blocked.

Under the least-congested-path policy, each node chooses the link that has the largest number of free wavelengths among all the alternate outgoing links. The set of free wavelengths consists of the set of wavelengths that are available on all the previous hops as well as the next outgoing link. Because each node only maintains local information on its outgoing links, no link state advertisement is required and thus the control overhead is greatly reduced. It has been shown that under light load, alternate-link routing outperforms both fixed routing and fixed-alternate routing. Under heavy load, fixed-alternate routing produces better performance.

> ***Example 5.2***: Consider the network in Example 5.1 with each link supporting four wavelengths, as shown in Figure 5.3. The wavelengths available on each link are indicated in the figure. Suppose that there is a connection request from node *a* to node *d*. Figure 5.3(a) illustrates the deflection routing for the connection request under the shortest-path-first policy. At node *a*, link *a-b* is chosen as the outgoing link because it leads to the shortest path to node *d* and the connection request is forwarded to node *b*. Similarly, at node *b*, link *b-c* is chosen as the outgoing link and the request is forwarded to node *c*. At node *c*, because there is no common wavelength available on links *a-b*, *b-c*, and *c-d*, link *c-e* is chosen as

the outgoing link because it leads to the second shortest path to node *d*. Accordingly, the connection request is deflected to node *e*, where it will proceed to node *d* on link *e-d*. As a result, a connection will be established for the connection request along route *a-b-c-e-d* on wavelength 1.

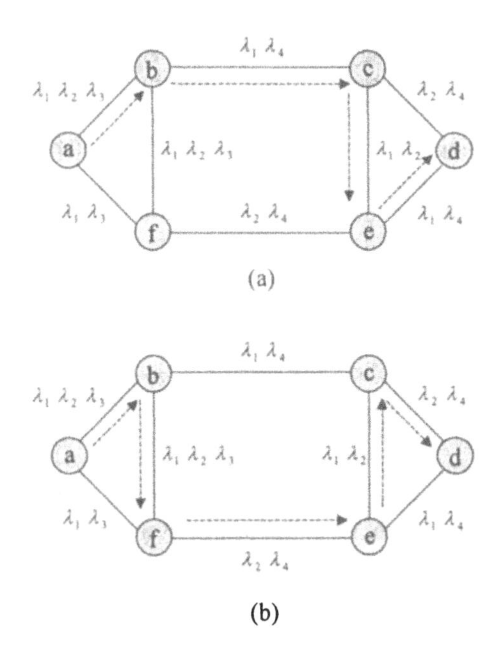

Figure 5.3 Illustration of deflection routing.

Figure 5.3(b) illustrates the deflection routing for the connection request under the least-congested-path-first policy. At node *a*, link *a-b* is chosen as the outgoing link because it has three wavelengths available whereas link *a-e* has only two wavelengths available. The connection request is then forwarded to node *b*. At node *b*, the connection request will be routed to node *f* as there are three wavelengths on link *b-f* whereas there is only one wavelength available on link *f-e*. The connection request will be routed in the same manner until it reaches node *d*. As a result, a connection will be established along route *a-b-f-e-c-d* on wavelength 2. In general, the least-congested-path policy results in longer paths than the shortest-path policy. However, it can route a connection around congested links and thus better balance the traffic load in the network. It has was in [6] that a shortest-path-first policy results in lower blocking under low traffic load, whereas a least-congested-path first policy results in lower blocking under higher traffic load.

5.3.3 Flooding-Based Routing

In flooding-based routing, a source node sends out a request packet on all its outgoing links. Each node receiving a request packet will forward the packet on all its outgoing links except the one corresponding to the incoming link on which the packet arrives. Before forwarding, the wavelength usage information on an outgoing link is collected and is included in the request packet that will be forwarded on that outgoing link. Each request packet may reach the destination node along a different route at a different time. When the destination node receives these request packets, it will decide on a route and select an available wavelength based on the collected information, and then send a reservation packet to the source node along the decided route to reserve the selected wavelength at each intermediate node.

The advantage of flooding-based routing is its simplicity in finding a route, in particular, a minimum delay route for a connection request in the current network state because it does not need any global information. However, flooding causes a huge number of control packets in the control channels, which would result in the control channels being congested and thus affect the network performance. One way to address this problem is to use a hop counter in the header of each packet and decrement it at each hop. When the counter of a packet reaches zero, the packet is discarded. The hop counter is usually initialized to a reasonable integer number, such as a number slightly larger than the number of hops on the shortest path between the source node and the destination node. Alternatively, we can also keep track of the flooded packets such that a duplicate packet is not forwarded. Here a duplicate packet is defined as a packet that has the same connection identification and does not carry any new wavelength information. For this purpose, when a node receives a request packet, it will check whether the packet carries a new connection identifier. If this is a new packet, the packet will be forwarded. Otherwise, the node will check whether the packet carries any new wavelength availability information. If it does, the packet will be forwarded. Otherwise, it will be discarded. A lightpath control protocol based on flooding-based routing will be presented in Section 5.5.5.

5.4 Wavelength Reservation

There are also two basic paradigms for wavelength reservation: parallel reservation and sequential reservation.

5.4.1 Parallel Reservation

In parallel reservation, the wavelength on each link along a decided route is reserved in parallel. Once a route is decided and a wavelength is selected, the source node sends a separate request packet to each node of the decided route. When each node receives a request packet, it will attempt to reserve the selected wavelength and then send either a positive or a negative acknowledgment packet back to the source node. If the source node receives a positive acknowledgment from all the nodes, a connection can be established successfully from the source node to the destination node. Otherwise, the connection request will be blocked. The advantage of parallel reservation is that it can reduce the connection setup time by having nodes to make wavelength reservation in parallel. However, because a separate request packet must be sent to each node of a decided route, parallel reservation requires more control overheads compared with sequential reservation. In addition, parallel reservation requires the source node to decide on an explicit route and select an available wavelength before making wavelength reservation. It usually assumes global information maintained at each network node, including network topology information and link state information.

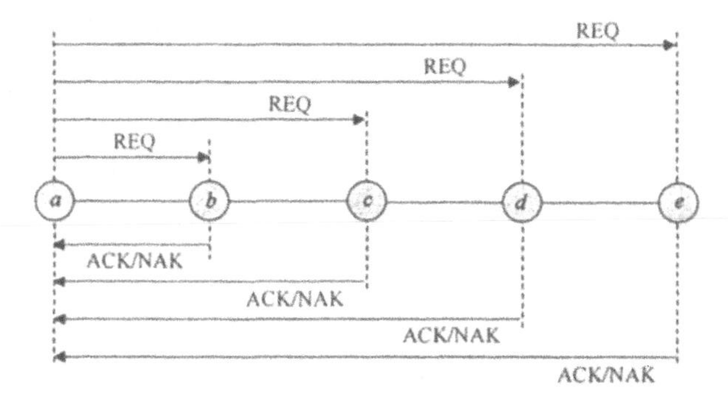

Figure 5.4 Illustration of parallel reservation.

Example 5.3: Consider a route consisting of five nodes with node *a* being the source node and node *e* the destination node, as shown in Figure 5.4. The route is decided by node *a* for a connection request to node *e*. An available wavelength is also selected at the same time. To reserve the selected wavelength on each link of the decided route in parallel, the source node first sends a separate request (*REQ*) packet to node *b*, node *c*, node *d*, and node *e*, respectively. In response, each of these nodes attempts to reserve the selected wavelength and then sends either a positive (*ACK*) or a negative (*NAK*) acknowledgment packet back to the source node, depending on the availability of the selected wavelength on each link. If node a receives an *ACK* packet from all the nodes, a connection can be established from node a to node e successfully for the connection request. Otherwise, the connection request will be blocked. It is obvious that four *REQ* packets and four *ACK* packets are needed to make the wavelength reservation.

An example that uses parallel reservation for lightpath establishment is the link state approach proposed in [1]. In this approach, each node in the network maintains global network state information about the network topology and wavelength usage on each link. A topology update protocol is employed at each node to periodically broadcast relevant topology information to all of the other nodes. On the arrival of a connection request, the originator first performs a route-computation algorithm to decide on a route and select an available wavelength based on the network state information it maintains. Once a route is decided and a wavelength is selected, the originator sends a *RESERVE* message in parallel to all the other nodes on the decided route to reserve the selected wavelength. In response, each of the nodes sends either a positive (i.e., a *RESERVE-ACK* message) or a negative (i.e., a *RESERVE-NACK* message) acknowledgment to the originator. If all the responses are positive, the reservation was successful at all the nodes. In this case, the originator will send a *SETUP* message to each of the nodes to configure the optical switches. If the reservation was not successful, the originator will send a *TAKEDOWN* message to each of the nodes to release the wavelength that was reserved earlier.

5.4.2 Sequential Reservation

In sequential reservation or hop-by-hop reservation, the wavelength on each link of a decided route is reserved on a hop-by-hop basis. Once a route is decided, the source node first sends a forward control packet to the destination node along the decided route. At each intermediate node, the control packet is processed and then forwarded to the next hop. When the destination node receives the control packet, it will send a backward control packet to the source node along the reverse route. The wavelength on each link of the decided route may be reserved at each intermediate node either by the forward control packet on its way to the destination node or by the backward control packet on its way to the source node, which is correspondingly referred to as forward reservation or backward reservation.

Example 5.4: Consider again the route used in Figure 5.5. To reserve a selected wavelength on each link of the decided route hop by hop, the source node first sends a forward request (*REQ*) packet to the destination node along the decided route. At each intermediate node, the *REQ* packet attempts to reserve the selected wavelength and then is forwarded to the next hop. When the destination node receives the *REQ* packet, it will send a backward acknowledgment (*ACK*) packet to the source node along the reserve route. When the source node receives the *ACK* packet, a connection has been established successfully from node *a* to node *e*. In contrast to parallel reservation, only one *REQ* packet and one *ACK* packet are needed to make the wavelength reservation.

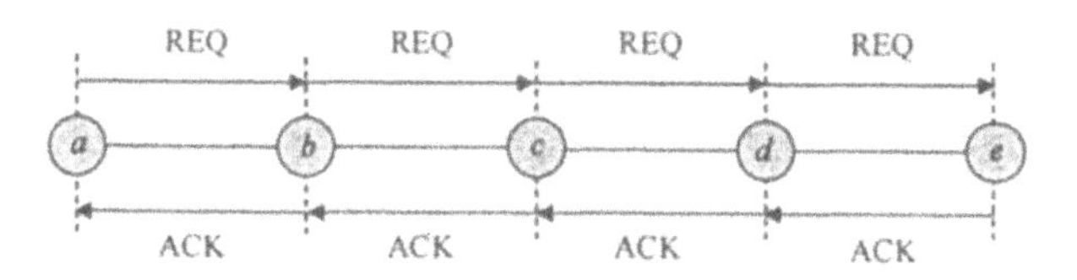

Figure 5.5 Illustration of hop-by-hop reservation.

Forward reservation

In forward reservation, also called source-initiated reservation (SIR), a wavelength is reserved by a forward control packet on its way from the source node (S-node) to the destination node (D-node) along a decided route. A basic forward reservation protocol (FRP) can be found in [10]. With FRP, the source node first sends a request (*REQ*) packet to the destination node along a decided route. At each intermediate node (I-node), the *REQ* packet will attempt to reserve a selected wavelength. If the wavelength can be reserved, the *REQ* packet will be forwarded to the next hop. Otherwise, a negative acknowledgment (*NAK*) packet will be sent back to the source node and the *REQ* packet will be dropped. The *NAK* packet will release all the wavelengths already reserved by the *REQ* packet and inform the source node of the reservation failure. If the *REQ* packet can reach the destination node, the destination node will send a positive acknowledgment (*ACK*) packet back to the source node along the reverse route. The *ACK* packet will configure the optical switch at each intermediate node. When the *ACK* packet reaches the source node, it implies that the lightpath has been established successfully. The signaling control with forward reservation is shown in Figure 5.6. The shaded area represents the period during which a wavelength is reserved but not in use. Obviously, a lot of bandwidth on the reserved wavelength is wasted during the reservation period, which would greatly decrease wavelength utilization. An effective way to address this problem is to use backward reservation.

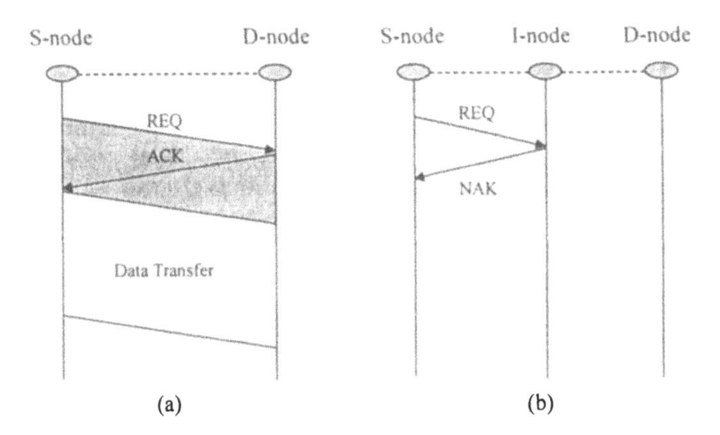

Figure 5.6 Forward reservation: (a) successful; (b) unsuccessful.

Backward reservation

In backward reservation, also called destination-initiated reservation (DIR), a wavelength is reserved by a backward control packet on its way from the destination node to the source node along the reverse of a decided route. A basic backward reservation protocol (BRP) can also be found in [10]. With BRP, the source node first sends a probe (*PROB*) packet to the destination node along a decided route. The *PROB* packet will not reserve any wavelength at each intermediate node. Instead, it will just collect wavelength availability information on each link along the decided route. When the destination node receives the *PROB* packet, it will select an available wavelength and then send a reservation (*RESV*) packet back to the source node along the reverse route. It is the *RESV* packet that reserves the selected wavelength and simultaneously configures the optical switch at each intermediate node on its way back to the source node. If the *RESV* packet cannot reserve the selected wavelength at an intermediate node, the node will send a negative acknowledgment (*NACK*) packet to the source node and the destination node. The *NACK* packet to the destination node will disconfigure the optical switches and release the wavelengths already reserved by the *RESV* packet, whereas the *NACK* packet to the source node will simply inform the source node of the reservation failure. If the *RESV* packet can reach the source node, it implies that the lightpath has been established successfully. Obviously, BRP can reduce bandwidth waste significantly compared with FRP, as shown in Figure 5.7.

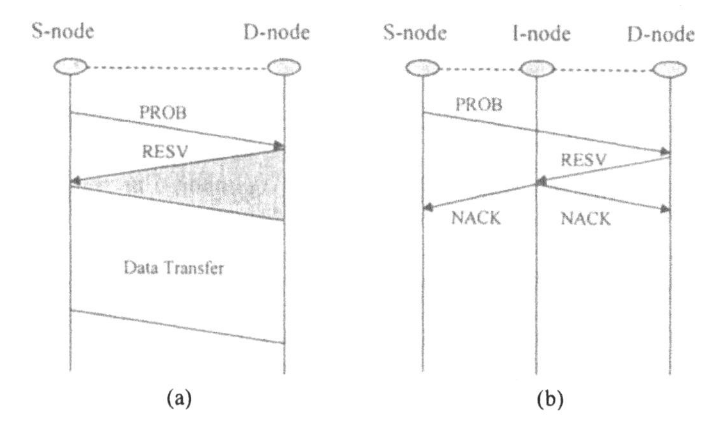

Figure 5.7 Backward reservation: (a) successful; (b) unsuccessful.

It should be pointed out that in either forward reservation or backward reservation, a reservation packet may not be able to reserve a selected wavelength at an intermediate node. This is because under dynamic traffic a selected wavelength that is available at one time may not be available at a later time. Because of the propagation delay on each fiber link, the link state information that was used to select a wavelength may be outdated, in particular, when a reservation packet reaches an intermediate node. As a result, a connection request is very likely to be blocked at the intermediate node because of the unavailability of the selected wavelength. To reduce the blocking probability due to outdated information, an intermediate-node-initiated reservation mechanism was proposed in [11] for networks with sparse wavelength conversion.

Intermediate-node-initiated reservation (IIR)

In backward reservation, when the reservation request reaches an intermediate node, the wavelength that was selected at the destination node may have already been reserved by another reservation request that arrived earlier. This will result in the corresponding connection request being blocked. It was shown in [11] that the delay between the instant at which the link state information at an intermediate node is collected and the instant at which the reservation request reaches the node has a significant impact on the blocking probability due to outdated information. A longer delay would lead to a larger blocking probability. For this reason, it is desirable to reduce this delay in order to reduce the blocking probability.

In a network with sparse wavelength conversion, wavelength converters are deployed at a subset of network nodes. Accordingly, a route can be divided into several segments whose end nodes can only be the source node, the destination node, or a node with wavelength converters. In this way, a lightpath does not have to use the same wavelength on each of the segments of a route. Instead, the wavelength reserved on one segment can be totally independent of the one reserved on another segment. In this case, the downstream end node of each segment can send a segment reservation request back to reserve a selected wavelength on the segment once it receives the connection request (i.e., the *PROB* packet). When the connection request reaches the destination node, a primary reservation request will be sent back to reserve a selected wavelength on the last segment and inform the source node of the lightpath setup status. By using this intermediate-node-initiated mechanism, the delay for reserving a wavelength on a link is reduced to the round-trip propagation delay between the corresponding node and the downstream end node of the segment. It was shown in [11] that the intermediate-node-initiated reservation mechanism can significantly reduce the overall blocking probability in the network. The following example further explains the concept of this reservation mechanism.

Example 5.3: Consider the route shown in Figure 5.8, where node 2 is a node with wavelength converters. To reserve an available wavelength with backward reservation, a *PROB* packet is sent to the destination node, which collects the wavelength availability information on each link of the route. When the destination node receives the *PROB* packet, it will select an available wavelength based on the collected information and send back a *RESV* packet to reserve the selected wavelength on each link. Note that, because of the wavelength conversion at node 2, the wavelength reserved on the first two links does not have to be the same as that reserved on link (2, 3). Accordingly, λ_1 can be reserved on link (0, 1) and link (1, 2), because both links have λ_1 and λ_2 available. Wavelength λ_3 can be reserved on link (2, 3) as both λ_3 and λ_4 are available on the link. However, because of the propagation delay on each link, the *RESV* packet may not be able to reserve the selected wavelength at each intermediate node. For example, when the wavelength availability information on link (0, 1) was collected at node 1 at time t_l, both λ_1 and λ_2 were available. However, when the *RESV* packet reaches node 1 at time $t_1 + \Delta t$ after a round-trip delay of Δt between node 1 and node 3, both λ_1 and λ_2 might be no longer available. This would result in the corresponding connection request being blocked.

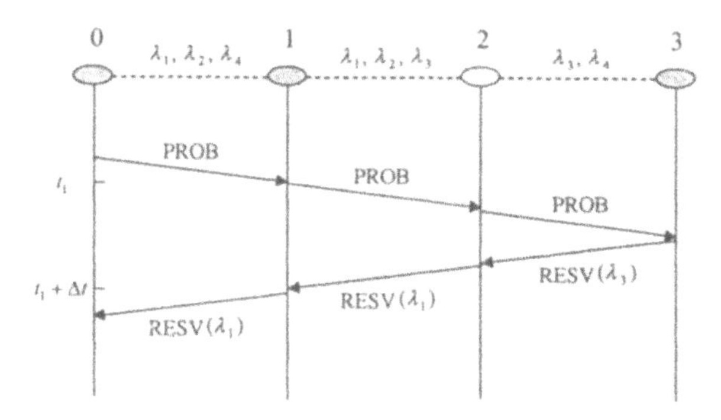

Figure 5.8 Illustration of backward reservation.

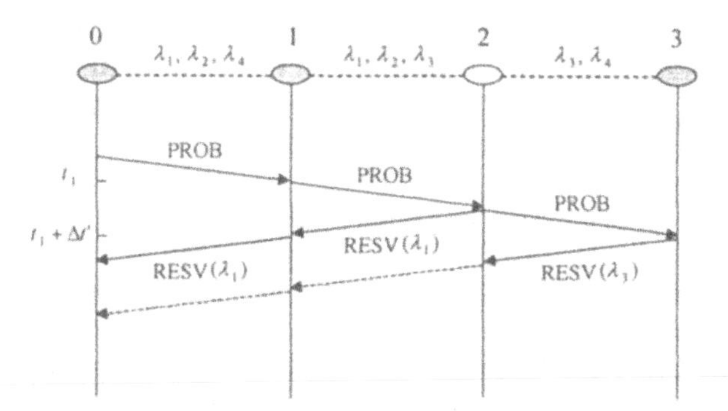

Figure 5.9 Illustration of intermediate-node-initiated reservation.

With immediate-node-initiated reservation, the route is divided into two segments. One includes link (0, 1) and link (1, 2), and the other includes link (2, 3). Once the *PROB* packet reaches node 2, the node can send a segment reservation request to reserve λ_1 on link (0, 1) and link (1, 2) of this segment. In this way, the reservation request will reach node 2 much earlier after a much shorter delay $\Delta t'$ than that in backward reservation, as shown in Figure 5.9. Accordingly, the probability that λ_1 becomes unavailable on link (0, 1) at time $t_l+\Delta t'$ might be largely reduced and λ_1 is more likely to be reserved.

5.4.3 Reservation Policies

During wavelength reservation, a reservation failure may occur at an intermediate node along a decided route. This is because under dynamic traffic a wavelength that is available at a node at the time a routing decision is made may become unavailable at a later time. In this case, there is a need for a reservation policy to handle such reservation failures. This reservation policy may have a big impact on the network performance in terms of request blocking probability and connection setup time. There are several policies already proposed in the literature, such as holding, aggressive, retrying, and one-way reservation.

Dropping versus Holding

The basic reservation protocols usually use a dropping policy. With the dropping policy, an intermediate node simply drops a reservation request in the event of a reservation failure and a negative acknowledgment packet is immediately sent back to the source node to inform of the reservation failure. Accordingly, the dropping policy is simple to implement but may result in a high blocking probability. To reduce the blocking probability, a holding policy can be introduced [10].

With the holding policy, an intermediate node does not drop a reservation request immediately in the event of a reservation failure. Instead, it will hold the request for a specified period of time to see whether the desired wavelength will be available during this period. If the wavelength becomes available, the wavelength will be reserved and the request forwarded. Otherwise, a negative acknowledgment packet will be sent back to the source node to inform of the reservation failure. Obviously, this can reduce the request blocking probability but would result in a longer connection setup time. It also increases the implementation complexity because each node must maintain a buffer to hold a reservation request and a timer to control the holding time. The holding policy can be used in both forward reservation and backward reservation.

Conservative versus Aggressive

The basic reservation protocols usually use a conservative policy. With the conservative policy, only one available wavelength is selected for reservation at a time. Because of dynamic traffic, there is no guarantee that the selected wavelength can be reserved on all links along a decided route. If the wavelength cannot be reserved, the source node may select a different wavelength for reservation at a later time. Obviously, this would result in a longer connection setup time because it may take several retries before a lightpath is established. To reduce the connection setup time, an aggressive policy can be introduced [10].

With the aggressive policy, more than one available wavelength is selected for reservation at a time on each link, with the expectation that at least one of the reserved wavelengths is available on all the links of a decided route. A greedy way is to reserve all possible wavelengths on each link. Specifically, the source node first reserves all possible wavelengths on the outgoing link and then sends a request packet that carries the reserved wavelengths to the next hop. At each intermediate node, a subset of wavelengths that consists of the common wavelengths available on both incoming link and outgoing link are computed and all the wavelengths in the subset are reserved. The request packet is then forwarded to the next hop carrying the subset of the reserved wavelengths. If the destination node receives the request packet, it will select one wavelength from the reserved wavelengths and send an acknowledgment packet back to the source node. The acknowledgment packet will release those wavelengths already reserved by the request packet but not selected on its way back. Obviously, this policy has a drawback, i.e., the over-reserved wavelengths during the reservation period, which would largely reduce wavelength utilization.

> ***Example 5.4***: Consider a physical route that consists of four nodes (i.e., node 0, node 1, node 2, and node 3), as shown in Figure 5.10. Suppose that forward reservation is used for wavelength reservation. During the reservation period, node 0 first reserves all available wavelengths on the first link, i.e., λ_1, λ_2, λ_3, λ_4, and then sends a reservation request to the next hop. At node 1, the common wavelengths on the incoming and outgoing links are reserved, i.e., λ_1, λ_3, λ_4, and the request is forwarded to node 2. Similarly, at node 2, the common wavelengths on the incoming and outgoing links are reserved, i.e., λ_3, λ_4, and the request is forwarded to node 3. When node 3 receives the request, one of the reserved wavelengths is selected, e.g., λ_3, and all the other reserved wavelengths will be released. Obviously, this policy is favorable to the current connection request. However, it is unfavorable to subsequent connection requests as it over-reserves the wavelength resources during the reservation period and the reserved wavelengths cannot be used by other connections.

A simple way to reduce the effect of over-reservation is to divide the wavelengths into several groups. Each node only reserves the wavelengths that belong to a particular group. The choice of a group is made at the source node and is based on the number of available wavelengths in each group. Usually, the source node will choose the group with the largest number of available wavelengths, and reserve all the available wavelengths in the group before sending the request to the next hop. The size of a group is a critical parameter. If the group is too large, too many wavelengths will be reserved. If the group is too small, the possibility of successful reservation will greatly decrease.

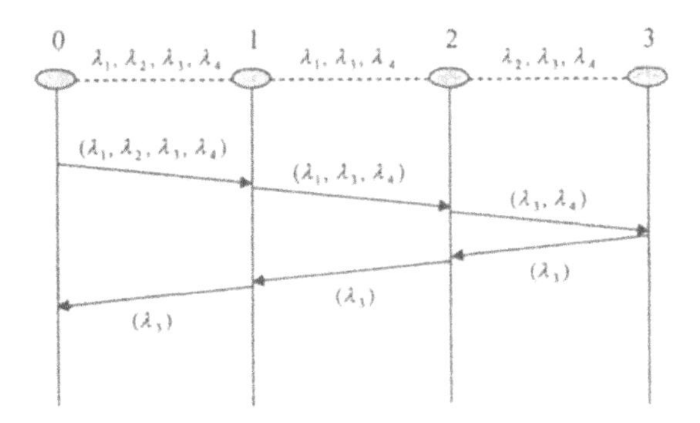

Figure 5.10 Illustration of aggressive reservation.

Nonretrying versus Retrying

In the basic backward reservation protocol, a negative acknowledgment packet is sent to the source node in the event of a reservation failure at an intermediate node and then the request is blocked. To increase the possibility of reservation success, a retrying policy has been proposed in [12–13]. With this policy, the destination node will select another available wavelength for reservation in the event of a reservation failure. Specifically, if there is a reservation failure at an intermediate node, a negative acknowledgment packet will be sent back to the destination node. When the destination node receives this message, it will select another available wavelength and send another reservation packet to the source node along the decided route, as shown in Figure 5.11. Obviously, this would increase the connection setup time. However, it has been shown that the retrying policy can significantly improve the request blocking probability without greatly increasing the setup delay as long as the number of retries is restricted to two or three [12].

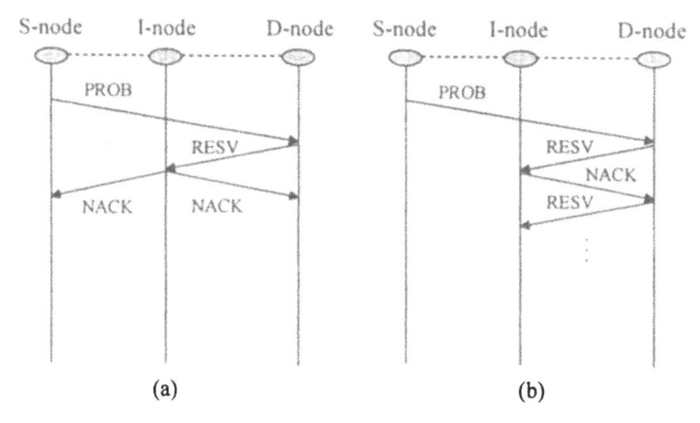

Figure 5.11 Illustration of retrying policy: (a) without retrying; (b) with retrying.

Two-Way versus One-Way

The basic reservation protocols usually use a two-way reservation policy. With the two-way reservation policy, the source node does not start data transfer until it receives an acknowledgment packet from the destination node. As a result, the connection setup time is at least twice as long as the propagation delay between the source node and the destination node. In large networks, this delay can be very large and thus is intolerable for some applications. To reduce this delay, a one-way reservation policy was proposed in [14].

With the one-way policy, the source node first sends out a request control packet to reserve a selected wavelength on each link of a decided route and at the same time to configure the optical switch at each intermediate node. Unlike with the two-way policy, the source node will not wait for an acknowledgment packet from the destination node. Instead, it will start data transfer before receiving an acknowledgement packet. To allow the control packet to have sufficient time to reserve the selected wavelength on each link and configure the optical switch at each intermediate node, data transfer will start T time units after the source node sends out the request packet. If the request packet cannot reserve the selected wavelength at an intermediate node, a negative acknowledgment packet will be sent back to the source node and the data will be dropped at the intermediate node. The source node can send out a request packet again at a later time. The difference between one-way reservation and two-way reservation is illustrated in Figure 5.12. Obviously, if one-way reservation succeeds, it can significantly reduce the connection setup delay as well as the bandwidth waste on the reserved wavelength during the reservation period. However, if the reservation fails, the data that are being transferred will be dropped, which would significantly affect wavelength utilization. To achieve good performance, the offset time T must be appropriately determined. Otherwise, data may catch up with the control packet, which would result in the failure of data transfer and thus reduce wavelength utilization.

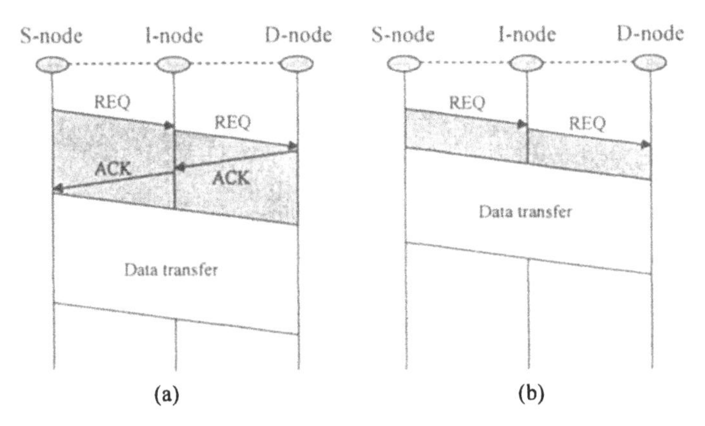

Figure 5.12 Illustration of two-way and one-way reservation: (a) two-way reservation; (b) one-way reservation.

5.5 Distributed Control Protocols for Lightpath Establishment

In this section, we present several typical distributed control protocols for lightpath establishment that have already been proposed in the literature.

5.5.1 Source Routing with Forward Reservation

In the source routing with forward reservation (SRFR) protocol, the source node is responsible for making a routing decision and forward reservation is used for wavelength reservation. The signaling control with this protocol can be briefly described as follows.

- Once the source node receives a connection request, it first performs a routing algorithm to decide on a route and select an available wavelength for the connection based on the network state information it maintains.

- Once a route is decided and an available wavelength is selected, the source node sends a request (*REQ*) packet to the destination node along the decided route, as shown in Figure 5.6.
- At each intermediate node, the *REQ* packet attempts to reserve the selected wavelength. If the wavelength can be reserved, the *REQ* packet will be forwarded to the next hop. Otherwise, a negative acknowledgment (*NAK*) packet will be sent back to the source node and the *REQ* packet will be dropped. The *NAK* packet will release all the wavelengths already reserved by the *REQ* packet and inform the source node of the reservation failure.
- If the *REQ* packet can reach the destination node, the destination node will send a positive acknowledgment (*ACK*) packet back to the source node along the reverse route. The *ACK* packet will configure the optical switch at each intermediate node.
- Once the *ACK* packet reaches the source node, it implies that the connection has been established successfully.

Note that the routing algorithm used in the SRFR protocol can be fixed routing, fixed-alternate routing, or adaptive routing, and the network state information used to make a routing decision can be either local or global information.

5.5.2 Source Routing with Backward Reservation

In the source routing with backward reservation (SRBR) protocol, the source node is responsible for making a routing decision and backward reservation is used for wavelength reservation. The signaling control with this protocol can be briefly described as follows.

- Once the source node receives a connection request, it first performs a routing algorithm to decide on a route for the connection based on the network state information it maintains.
- Once a route is decided, the source node sends a probe (*PROB*) packet to the destination node along the decided route, as shown in Figure 5.7. The *PROB* packet does not reserve any wavelength at each intermediate node. Instead, it just collects wavelength availability information on each link along the decided route.
- Once the destination node receives the *PROB* packet, it selects an available wavelength based on the collected information and then sends a reservation (*RESV*) packet back to the source node along the reverse route. It is the *RESV* packet that attempts to reserve the selected wavelength and simultaneously configure the optical switch at each intermediate node on its way back to the source node.

- If the *RESV* packet cannot reserve the selected wavelength at an intermediate node, the node will send a negative acknowledgment (*NACK*) packet to the source node and the destination node. The *NACK* packet to the destination node will disconfigure the optical switches and release the wavelengths already reserved by the *RESV* packet whereas the *NACK* packet to the source node will simply inform the source node of the reservation failure.
- If the *RESV* packet can reach the source node, it implies that the connection has been established successfully.

Similar to SRFR, there is no restriction on the routing algorithm used in the SRBR protocol and the source node can use either local or global information to make a routing decision.

5.5.3 Destination Routing with Backward Reservation

In the destination routing with backward reservation (DRBR) protocol [3], the destination node is responsible for making a routing decision based on the global network state information it maintains, which is updated either periodically or on a change. The signaling control with this protocol can be briefly described as follows.

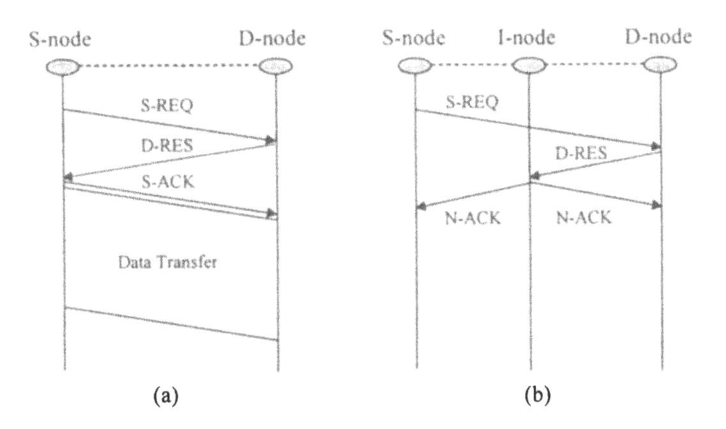

Figure 5.13 Signaling control with DRBR.

- Once the source node receives a connection request, it first sends a request (*S-REQ*) packet to the destination node along the shortest route, as shown in Figure 5.13. This *S-REQ* packet does not do anything on its way to the destination node.

- Once the destination node receives the *S-REQ* packet, it first performs a routing algorithm to decide on a route and select an available wavelength for the connection based on the global network state information it maintains. If no route is available, the destination node will simply send a negative acknowledgment (*N-ACK*) packet to the source node to inform of the failure. If there is a route available, the destination node will send a reservation (*D-RES*) packet to the source node along the reverse route, which carries both the route and wavelength information.
- At each intermediate node, the *D-RES* packet attempts to reserve the selected wavelength and simultaneously configure the optical switch.
- If the selected wavelength can be reserved, the *D-RES* packet will be forwarded to the next hop. Otherwise, an *N-ACK* packet will be sent to the source node and the destination node. The *N-ACK* packet to the destination node will disconfigure the optical switches and release the wavelength already reserved by the *D-RES* packet on its way to the destination node, whereas the *N-ACK* packet to the source node will simply inform the source node of the reservation failure.
- If the *D-RES* packet reaches the source node, it implies that the connection has been established successfully. In this case, the source node will send an acknowledgment packet (*S-ACK*) to the destination node to inform of the reservation success.

With DRBR, the destination node makes a routing decision after it receives an *S-REQ* packet from the source node. This makes it possible to use more recent network state information to make a routing decision compared with source routing and can therefore significantly reduce the possibility of a reservation failure. On the other hand, because the *S-REQ* packet is delivered in the control channels and does not need to do anything on its way to the destination node, it can always take the shortest-path route to the destination node, which could reduce the connection setup time to a certain extent. Note that DRBR is based on global network state information maintained at the destination node. This requires link state advertisement in the network, which would largely increase the control overhead in the network.

5.5.4 Alternate-Link Routing with Backward Reservation

In the alternate-link routing with backward reservation (ARBR) protocol, each node employs an alternate-link routing algorithm [6] to decide the next hop, as described in Section 5.3.2. The signaling control with this protocol can be briefly described as follows.

- Once the source node receives a connection request, it chooses one link from a set of alternate outgoing links to the destination node based on some criterion, such as the shortest-path first or the least-congested-link first, and then sends the request packet to the next hop. This packet carries the connection identifier, the destination address, a set of free wavelengths on the outgoing link, and an empty path list.
- Each intermediate node independently chooses one link from a set of alternate links based on the same criterion and then forwards the request packet to the next hop. Before it forwards the request, it intersects the set of free wavelengths carried in the request packet with the set of free wavelengths on the outgoing link and then modifies the set of free wavelengths and path list carried in the request packet.
- Once the destination node receives the request packet, it selects a free wavelength in the set of free wavelengths contained in the packet, configures the local OXC, and then sends a reservation packet to the source node along the path experienced by the request packet.
- At each upstream node, the reservation packet attempts to reserve the selected wavelength. If the wavelength can be reserved, the node will configure the local OXC and forward the reservation packet to the upstream node until the reservation packet is received at the source node.

Note that because each node only maintains local information on its outgoing links, no link state advertisement is required and thus no additional control overhead is introduced.

5.5.5 Flooding-Based Routing with Backward Reservation

In the flooding-based routing with backward reservation (FRBR) protocol [15], flooding-based routing is used and the destination node is responsible for making a routing decision. Backward reservation is used for wavelength reservation. The signaling control with this protocol can be briefly described as follows.

- Once the source node receives a connection request, it first sends a probe (*PROB*) packet to all or some of its neighbors. The *PROB* packet originally contains the connection identifier, the destination address, a set of free wavelengths on an outgoing link, and an empty path list. If the set of free wavelengths on an outgoing link is empty, the *PROB* message will not be sent to that outgoing link.
- Once an intermediate node receives a *PROB* packet, it forwards the packet to all or some of its neighbors. Before it forwards the packet to a neighbor, the node will first check whether the *PROB* packet has passed

through it before. If it is not the first time the packet passes through this node, the *PROB* packet will be dropped. Otherwise, the node will intersect the set of free wavelengths contained in the packet with the set of free wavelengths on the outgoing link. If the subset is empty, the *PROB* packet will not be forwarded onto the outgoing link. Otherwise, the node will modify the set of free wavelengths and the path list contained in the *PROB* packet and then forward the *PROB* packet onto the outgoing link.

- Once the destination node receives a *PROB* packet, it selects a free wavelength from the set of free wavelengths contained in the packet, configures the local optical switch, and sends a reservation (*RESV*) packet to the source node along the path specified in the path list.
- At each upstream node, the *RESV* packet attempts to reserve the selected wavelength. If the wavelength can be reserved, the node will configure the local optical switch and then forward the *RESV* packet to the upstream node until the packet is received at the source node. If the wavelength cannot be reserved, a negative acknowledgment (*NACK*) packet will be sent to the destination node, which will disconfigure the optical switch and release the wavelength already reserved by the *RESV* packet at each intermediate node.
- Once the destination node receives the *NACK* packet, and if it has already received a *PROB* packet on another path, the destination node will send another *RESV* packet to the source node along that path. If the source node receives the *RESV* packet, it implies that the connection has been established successfully.

Note that the destination node may receive more than one *PROB* packet, each arriving at a different time and containing a different path list. To achieve the smallest delay, the destination node chooses the path found by the *PROB* packet that arrives first. If a second *PROB* packet arrives for the same connection identifier, the destination node will store the packet until the upstream reservation is confirmed. If the reservation is not successful, the destination node will choose the second stored path. For more details about this protocol, the readers are referred to [15].

5.5.6 Multiple-Path Routing with Backward Reservation

The multiple-path routing with backward reservation (MRBR) protocol is a variation of the FRBR protocol. In the MRBR protocol [15], each node maintains a routing table that contains a set of fixed-alternate routes to each destination node. The destination node is responsible for making a routing decision, and backward reservation is used for wavelength reservation. The signaling control with this protocol can be briefly described as follows.

- Once the source node receives a connection request, it sends a probe (*PROB*) packet to the destination node along each of the alternate routes. The *PROB* packet originally contains the connection identifier, the destination address, a set of free wavelengths on an outgoing link, and a path identifier.
- Once an intermediate node receives a *PROB* packet, it intersects the set of free wavelengths contained in the packet with the set of free wavelengths on the outgoing link and forwards the packet to the next hop.
- Once the destination node receives a *PROB* packet, it selects a free wavelength from the set of free wavelengths contained in the packet, configures the local optical switch, and sends a reservation (*RESV*) packet to the upstream node along the path specified in the path identifier.
- At each upstream node, the *RESV* packet attempts to reserve the selected wavelength. If the wavelength can be reserved, the node will configure the local optical switch and then forward the *RESV* packet to the upstream node until the packet is received by the source node. If the wavelength cannot be reserved, a negative acknowledgment (*NACK*) packet will be sent to the destination node, which will disconfigure the optical switches and release the wavelength already reserved by the *RESV* packet at each intermediate node.
- Once the destination node receives the *NACK* packet, and if it has already received a *PROB* packet on another path, the destination node will send another *RESV* packet to the source node along that path. If the source node receives the *RESV* packet, it implies that the connection has been established successfully.

It was shown in [15] that the MRBR protocol can provide a shorter connection setup time than the FRBR protocol but at the cost of the blocking probability.

5.5.7 Neighborhood Information-Based Routing with Backward Reservation

In the neighborhood information-based routing with backward reservation (NRBR) protocol, neighborhood information-based routing [2] and backward reservation are employed to establish a lightpath for each connection request. In neighborhood information-based routing, each node maintains a routing table that stores a set of precomputed link-disjoint alternate routes to each of the other network nodes. The source node of a connection request maintains the wavelength usage information on the first k links of all the routes to the destination node, which is collected on demand or exchanged periodically. The source node uses such neighborhood information to make a routing decision for the connection request. The signaling control with this protocol can be briefly described as follows.

- On the arrival of a connection request, the source node sends a probe (*PROB*) packet to the destination node along each of the alternate routes. At each intermediate node, the *PROB* packet collects the wavelength usage information and is then forwarded to the next hop. When the *k*th node on a route receives the *PROB* packet, it sends an acknowledgment (*ACK*) packet back to the source, containing a set of wavelengths available on all *k* links. When the source node receives all the *ACK* packets from all the *k*th nodes, it will perform a routing algorithm to decide on a route among all the alternate routes based on the collected neighborhood information. Here the neighborhood information is collected on demand. Alternatively, the neighborhood information can also be updated periodically. In this case, on the arrival of a connection request, the source node will perform a routing algorithm to make a routing decision based on the neighborhood information it maintains. In both cases, if there is no wavelength available on the first *k* links of any of the alternate routes, the connection request will be blocked.
- Once a route is decided, the source node employs the backward reservation protocol to reserve a wavelength along the decided route. Specifically, the source node first sends a probe (*PROB*) packet to the destination node along the decided route. At each intermediate node, the *PROB* packet collects the wavelength usage information and is then forwarded to the next hop. When the destination node receives the *PROB* packet, it selects an available wavelength based on the collected information and then sends a reservation (*RESV*) packet to reserve the selected wavelength at each intermediate node along the reserve route. If there is no wavelength available on the decided route, the connection request will be blocked.

Note that the source node can also employ forward reservation to reserve a wavelength along the decided route.

5.6 Summary

Lightpath control is one of the most important problems for achieving good network performance in wavelength-routed WDM networks. Distributed lightpath control is considered more suitable than centralized control for large networks with dynamic traffic. This chapter discussed distributed lightpath control with a focus on lightpath establishment. The main problems with distributed lightpath establishment were discussed, including routing, wavelength assignment, and wavelength reservation. In terms of routing, there are three basic routing paradigms: explicit routing, hop-by-hop routing, and flooding-based routing. Both explicit routing and hop-by-hop routing can be based on local or global network state information. Explicit routing can further be classified into three basic paradigms: fixed routing, fixed-alternate routing, and adaptive routing. In general, fixed routing is the simplest paradigm to implement but may result in a higher blocking probability. Adaptive routing can significantly reduce the blocking probability but has a higher computational complexity. Fixed-alternate routing provides a tradeoff between computational complexity and network performance. In terms of wavelength reservation, there are two basic reservation paradigms: parallel reservation and hop-by-hop reservation. In general, parallel reservation can reduce the reservation time but would cause more control overhead whereas hop-by-hop reservation can reduce control overhead but would increase the reservation time. Hop-by-hop reservation can be further classified into two basic paradigms: forward reservation and backward reservation. It has been shown that backward reservation performs better than forward reservation in terms of wavelength utilization and blocking probability. In addition, an intermediate-node-initiated reservation mechanism was also introduced, which can be used to improve wavelength utilization in networks with sparse wavelength conversion. Based on the basic routing and wavelength reservation paradigms, several typical distributed control protocols for lightpath establishment were also presented.

Problems

5.1 Why is distributed control preferred to centralized control for large networks with dynamic traffic?

5.2 What are the major issues involved in distributed lightpath control? Do they have any differences from those in centralized lightpath control?

5.3 What are the major concerns with routing under distributed control?

5.4 Compare the advantages and disadvantages of using local, global, and neighborhood information in routing.

5.5 What are the major concerns with wavelength reservation under distributed control?

5.6 Compare the advantages and disadvantages of explicit routing and hop-by-hop routing.

5.7 Explain why flooding-based routing can find the minimum delay route. How flooded control packets be controlled in flooding-based routing?

5.8 Compare the performance of forward reservation and backward reservation. Why does backward reservation generally perform better than forward reservation in terms of wavelength utilization?

5.9 How does a wavelength reservation failure occur? How does it affect network performance?

5.10 Explain the advantages and disadvantages of various reservation policies, including dropping and holding, conservative and aggressive, nonretrying and retrying, and two-way and one-way.

References

[1] Rajiv Ramaswami and Adrian Segall, "Distributed network control for optical networks," *IEEE/ACM Transactions on Networking,* vol. 5, no. 6, Dec. 1997, pp. 936–943.

[2] Ling Li and Arun K. Somani, "Dynamic wavelength routing using congestion and neighborhood information," *IEEE/ACM Transactions on Networking,* vol. 7, no. 5, Oct. 1999, pp. 779–786.

[3] Jun Zheng and Hussein T. Mouftah, "Distributed lightpath control based on destination routing for wavelength-routed WDM networks," *SPIE Optical. Networks. Magazine,* vol. 3, no. 4, Jul./Aug. 2002, pp. 38–46.

[4] S. Ramamurthy and B. Mukherjee, "Fixed-alternate routing and wavelength conversion in wavelength-routed optical networks," *Proc. of IEEE GLOBECOM'98*, vol. 4, Nov. 1998, pp. 2295–2302.

[5] H. Harai, M. Murata, and H. Miyahara, "Performance of alternate routing methods in all-optical switching networks," *Proceedings of IEEE INFOCOM'97*, vol. 2, Kobe, Japan, Apr. 1997, pp. 516–524.

[6] Jason P. Jue and Gaoxi Xiao, "An adaptive routing algorithm for wavelength-routed optical networks with a distributed control scheme," *Proceedings of 9th International Conference on Computer Communications and Networks (IC3N'00),* Las Vegas, Nevada, Oct. 2000, pp. 192–197.

[7] Ching-Fang Hsu, Te-Lung Liu, Nen-Fu Huang, "On adaptive routing in wavelength-routed networks," *SPIE Optical Networks Magazine*, vol. 3, no. 1, Jan./Feb. 2002, pp. 15–24.

[8] Kit-man Chan and Tak-shing Perter Yum, "Analysis of least congested path routing in WDM lightwave networks," *Proc. of IEEE INFOCOM'94*, vol. 2, Toronto, Canada, Apr. 1994, pp. 962–969.

[9] Hui Zang et al., "Connection management for wavelength-routed WDM networks," *Proceedings of IEEE CLOBECOM'99,* Rio de Janeiro, Brazil, Dec. 1999, pp. 1428–1432.

[10] Xin Yuan, Rami Melhem, and Rajiv Gupta, "Distributed path reservation algorithms for multiplexed all-optical interconnection networks," *IEEE Transactions on Computer*, vol. 48, no. 12, Dec. 1999, pp. 1355–1363.

[11] Kejie Lu et al., "Intermediate-node initiated reservation (IIR): A new signaling scheme for wavelength-routed WDM networks," *IEEE Journal on Selected Areas on Communications*, vol. 21, no. 8, Oct. 2003, pp. 1231–1240.

[12] D. Saha, "An efficient wavelength reservation protocol for lightpath establishment in all-optical networks (AONs)," *Proc. of IEEE Globecom'00*, vol. 2, San Francisco, Nov. 2000, pp. 1264–1268.

[13] Debashis Saha, "A comparative study of distributed protocols for wavelength reservation in WDM optical networks," *SPIE Optical Networks Magazine,* vol. 3, no. 1, Jan./Feb. 2002, pp. 45 52.

[14] Yousong Mei and Chunming Qiao, "Distributed control schemes for dynamic lightpath establishment in WDM optical networks," Proceedings of Optical Networks Workshop, Richardson, Texas, Jan./Feb. 2000.

[15] Abdallah A. Shami et al., "Integrated routing/signaling protocols in GMPLS-based optical networks," *SPIE Optical Networks Magazine*, vol. 3, no. 4, Jul./Aug. 2002, pp.16–23.

Chapter 6

Optical Layer Survivability

6.1 Introduction

Network survivability has been a great concern in all types of high-speed telecommunications networks [1]. In wavelength-routed WDM networks, an optical fiber supports a number of optical channels or wavelengths, each operating at a very high rate of several gigabits per second. Although this provides a huge transmission bandwidth to carry data traffic, it also introduces a potential problem. Because of the huge transmission bandwidth, a single network failure such as a fiber cut may cause a large amount of data loss in the network, which would largely degrade and even disrupt network services. For this reason, network survivability is of particular importance in such networks. To guarantee network services, it is imperative to incorporate effective protection and restoration capabilities in the network to provide a high level of service survivability against different types of network failures, such as a fiber cut or a node fault. From the viewpoint of layered architecture, survivability can be provided at different network layers, such as IP, ATM, SONET, and the optical layer. Although each of the higher layers may have its own protection and restoration mechanisms, it is still attractive to provide survivability at the optical layer because of a number of advantages, such as fast service recovery, efficient resource utilization, and protocol transparency [2–3]. For this reason, optical layer survivability has received a lot of attention in the design of wavelength-routed WDM networks [2–8].

In this chapter, we discuss optical layer survivability in wavelength-routed WDM networks. The need for optical layer survivability is first explained, and the fundamental concepts of protection and restoration for optical layer survivability are introduced. A variety of basic protection and restoration schemes are also introduced, and both the survivable network design for static traffic and survivable routing for dynamic traffic are discussed. For static traffic, the survivable network design problem can be formulated as a mixed-integer linear programming (MILP) problem. Two examples of ILP formulations are presented, which are based on dedicated path protection and shared path protection, respectively. For dynamic traffic, several important survivable routing algorithms already proposed in the literature are presented. Moreover, dynamic restoration is discussed and several dynamic restoration protocols for surviving single-link failures are presented.

6.2 Need for Optical Layer Survivability

In wavelength-routed WDM networks, data traffic is transferred over all-optical connections or lightpaths, each operating at a very high rate of a few gigabits per second (e.g., 2.5 Gbps or 10 Gbps). These lightpaths constitute an optical layer between the physical layer and the higher layers (as described in Section 1.3.2), which provides transmission bandwidth to its higher layers. The higher layers make use of the transmission bandwidth provided by the optical layer to provide different kinds of network services. With WDM technology, a single optical fiber provides a number of optical channels or wavelengths, which can carry a huge amount of data traffic. As mentioned in Section 1.1, commercially available WDM systems can support up to 160 OC-192 (10 Gbps) channels on a single fiber, which is equivalent to a bandwidth of 1.6 TBps. In a multiple-fiber network, multiple fibers are deployed on each fiber link, which further multiplies the transmission bandwidth in the network.

Although a wavelength-routed WDM network has a huge transmission bandwidth to carry data traffic, it also has a potential problem. Because of the huge transmission bandwidth inherent in each fiber, a single network failure may lead to a large amount of data loss and thus result in severe service degradation and disruption. For example, a fiber cut can cause the disruption of all the lighpaths that traverse the failed fiber, which would result in a large amount of data loss and thus largely degrade and even disrupt the network services. In view of the fact that on an average a fiber cable carries about 100 fibers, a cable cut may result in the data loss of several terabits per second. For this reason, it is imperative to incorporate effective protection and restoration capabilities in the network to provide a high level of service survivability against different types of network failures, such as a fiber cut or a node fault. Network survivability refers to the capability of a network to recover from network failures and maintain a high level of service availability. A network with survivability capability is referred to as a survivable network.

In wavelength-routed WDM networks, there are two basic types of network failures: link failure and node failure, which are common in all types of telecommunications networks. A link failure is usually caused by a fiber or cable cut, which is considered one of the most common failures in optical networks and can result from a variety of reasons, such as natural phenomena (e.g., flood and earthquake) or human errors [9]. A node failure may be caused by equipment failures at network nodes because of natural phenomena, human errors, and hardware degradation [9]. A node failure, although not as common as a link failure, may affect more network users than a link failure. For example, if an optical switch breaks down, all the fibers connected to the switch will also fail, resulting in the disruption of all the connections that are carried in all the fibers. In addition to the link and node failures, there is another type of network failures, called channel failures, which may be caused by the failure of an equipment component associated with a particular channel, such as a transmitter or a receiver.

According to layered architecture, network survivability can be provided at different network layers, including the optical layer and the higher layers. Network survivability can be achieved through various protection and restoration mechanisms. Although each of the higher layers may have its own protection and restoration mechanisms, it is still attractive to provide effective protection and restoration mechanisms at the optical layer because of the following main reasons [2–3]:

- The optical layer can provide protection and restoration functions that the higher layers may not be able to provide.
- The optical layer can provide faster service recovery (on the order of milliseconds) than the higher layers (on the order of seconds).
- The optical layer can provide protection and restoration against network failures more efficiently than the higher layers.
- The optical layer can provide an additional level of survivability in the network, such as against multiple failures.

- The optical layer can provide protection and restoration with significant cost savings.

Despite the above advantages, optical layer protection and restoration also have some limitations [2–3]. For example, because the optical layer is protocol transparent, it is unable to detect increased bit error rates of traffic carried in a lightpath. Also, it is unable to handle traffic at a granularity finer than one lightpath.

6.3 Protection and Restoration for Optical Layer Survivability

In this section, we introduce the fundamental concepts of protection and restoration for optical layer survivability.

At the optical layer, a network failure could be a link failure, a node failure, or a channel failure, as mentioned in Section 6.2. Usually, a link failure occurs much more frequently than a node or channel failure. This is because a fiber cable usually extends over a very long distance under various geographical conditions, such as in mountains, under oceans, or in remote areas. It is more vulnerable to various natural phenomena (such as flood and earthquake) and human errors and is more difficult to monitor and maintain. In contrast, a network node is usually installed in a central office, which is easier to monitor and maintain. Moreover, the switching equipment at a node usually has redundancy for key components, such as the switch fabric and the control unit, in order to maintain a high level of reliability. As a result, a node or channel failure is much less likely to occur than a link failure. Even though a node failure occurs, it is much easier and faster to locate and repair the failure. For this reasons, link failures have received much more attention than node failures.

On the other hand, single failures have been studied more extensively than multiple failures [4][10]. In fact, most network providers do not plan for protection and restoration against multiple simultaneous failures because it would be prohibitively costly to protect against such failures, which have a very low probability occurring in practice [10]. This actually assumes that the networks are designed well enough that multiple simultaneous failures occur very rarely or in other words, it is much less likely that a new failure will occur before the recovery of an old failure. In view of all the above facts, we will focus on protection and restoration against single-link failures unless otherwise stated.

6.3.1 Protection for Point-to-Point WDM Links

The automatic protection switching (APS) in SONET systems can be used in point-to-point WDM links. There are two basic types of APS schemes for protection against link failures: 1+1 protection and 1:1 protection, or more generally, 1:N protection, as shown in Figure 6.1. The concept of APS can also be applied to protection and restoration in wavelength-routed WDM networks.

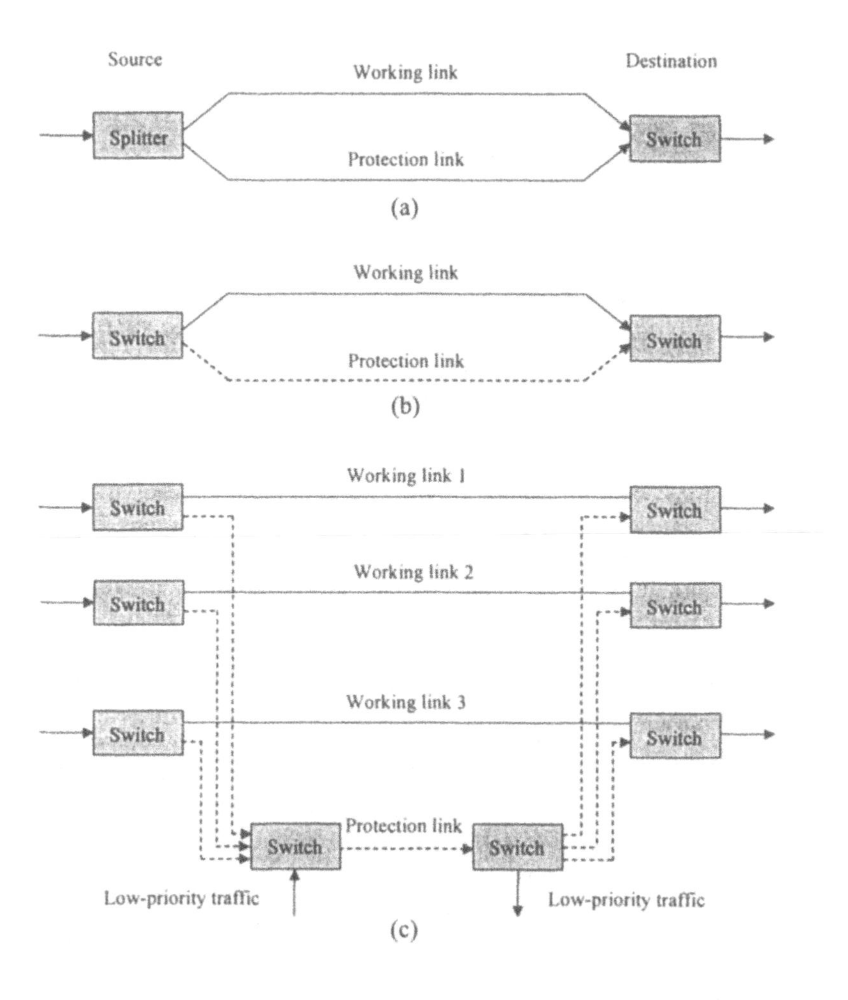

Figure 6.1 Automatic protection switching: (a) 1+1 protection; (b) 1:1 protection; (c) 1:N protection.

1+1 Protection

In 1+1 protection, an optical signal is transmitted simultaneously on two fiber links between the transmitting end and the receiving end. One of the links is referred to as the primary link or working link, and the other is referred to as the backup link or protection link. The receiving end selects one of the links for receiving the signal, which is usually the link with a better signal quality. If the working link fails, the receiver only needs to simply switch over to the protection link and then continues to receive the signal.

The advantage of 1+1 protection is that it can provide very fast service recovery. However, the transmitting end usually uses a splitter to split up a single signal onto two fiber links, which may result in a 3-dB additional signal loss at the transmitting end.

1:1 Protection

In 1:1 protection, there are still two fiber links between the transmitting end and the receiving end, one being the working link and the other the protection link. However, an optical signal is transmitted over only one link, i.e., the working link. If the working link fails, both the transmitter and the receiver need to switch over to the protection link.

1:1 Protection has two main advantages over 1+1 protection. The first is that in the absence of a link failure the protection link can be used to transmit a signal that carries lower-priority traffic, which, to a certain extent, can increase link utilization. If the working link fails, the signal on the protection link must be terminated and the protection link is used to transmit the signal transmitted on the working link. The second advantage is that it is possible to share a single protection link among multiple working links, which leads to a more general form, 1:N protection. However, in unidirectional communication systems, 1:1 protection is not as fast as 1+1 protection in service recovery. In such systems, a signal is transmitted over a fiber link in only one direction. If the working link fails, only the receiving end can detect the link failure. In this case, the receiving end must use a signaling protocol to notify the transmitting end of the failed link so that the transmitter can switch over to the protection link. For this reason, 1:1 protection is not as fast as 1+1 protection in unidirectional communication systems.

1:N Protection

In 1:N protection, a single protection link is shared among multiple working links. It can handle the failure of any single working link. If more than one working links fail, only the signal on one of the working links can be switched over to the protection link. It is obvious that 1:N protection can increase the utilization of the protection link. However, it is unable to handle multiple link failures that occur simultaneously.

6.3.2 Static Protection and Dynamic Restoration

Network survivability at the optical layer is based on two basic survivability paradigms: static protection and dynamic restoration.

Static protection

In static protection, also referred to as predetermined protection, spare network resources are reserved during network design or at the time of connection establishment for protection against network failures. In the event of a network failure, the disrupted network services are recovered by using the reserved network resources. Obviously, static protection is fast in service recovery as the reserved network resources are dedicated for network failures. However, it is inefficient in resource utilization because the reserved network resources cannot be used for other traffic demands.

Dynamic restoration

In dynamic restoration, no spare network resources are reserved during network design or at the time of connection establishment. The network must search dynamically for spare network resources available in the network to recover the disrupted network services after network failures occur. Accordingly, dynamic restoration is more efficient in resource utilization than static protection as no network resources are reserved before any network failure occurs. However, it is slower in service recovery because it takes time to search dynamically for spare network resources. Moreover, the service recovery cannot be one hundred percent guaranteed because the network may not have sufficient spare resources at the occurrence of network failures.

Although both static protection and dynamic restoration have their own advantages and disadvantages, most of today's optical networks are still using static protection mechanisms rather than dynamic restoration mechanisms. For example, APS has been widely used in point-to-point WDM links. Self-healing rings (SHRs) have been used in WDM ring networks. For a large mesh network, such protection mechanisms may require a large amount of spare capacity and thus largely reduce resource utilization in the network. For this reason, dynamic restoration becomes attractive. To have a trade-off between service recovery and resource utilization, a practical strategy is to use static protection as a primary mechanism to guarantee service recovery, while using dynamic restoration as an auxiliary mechanism to increase resource utilization. It should be pointed out that the terms "protection" and "restoration" are often used interchangeably in optical networks. For clarity of exposition, the terms "protection" and "restoration" used in this chapter specifically refer to static protection and dynamic restoration, respectively.

6.3.3 Dedicated Protection and Shared Protection

From the perspective of resource sharing, static protection can be further classified into two basic paradigms: dedicated protection and shared protection.

Dedicated protection

In dedicated protection, the reserved network resources are dedicated to each single failure. A typical example of dedicated protection is 1+1 protection or 1:1 protection (if the reserved network resources are not used to deliver low-priority traffic). Obviously, dedicated protection is very fast in service recovery and can guarantee service recovery from any failure. However, it is inefficient in resource utilization because it requires one hundred percent spare capacity reserved. Under dedicated protection, the best network resource utilization is only fifty percent.

Shared protection

In shared protection, the reserved network resources can be shared among multiple failures. A typical example of shared protection is 1:N protection. Because of the resource sharing, shared protection is more efficient in resource utilization and can thus result in higher network utilization. In wavelength-routed WDM networks, shared protection provides the same level of protection against single path failures as dedicated protection, with potentially higher network utilization [11]. However, shared protection cannot handle multiple failures that share the same reserved network resources and occur simultaneously.

In practice, dedicated protection and shared protection can complement each other to provide more flexible protection solutions. For example, dedicated protection can be used for those paths that require a high level of protection while the other paths can be protected under shared protection in order to achieve high network utilization.

6.3.4 Link Protection and Path Protection

From another perspective, static protection can also be classified into link protection and path protection.

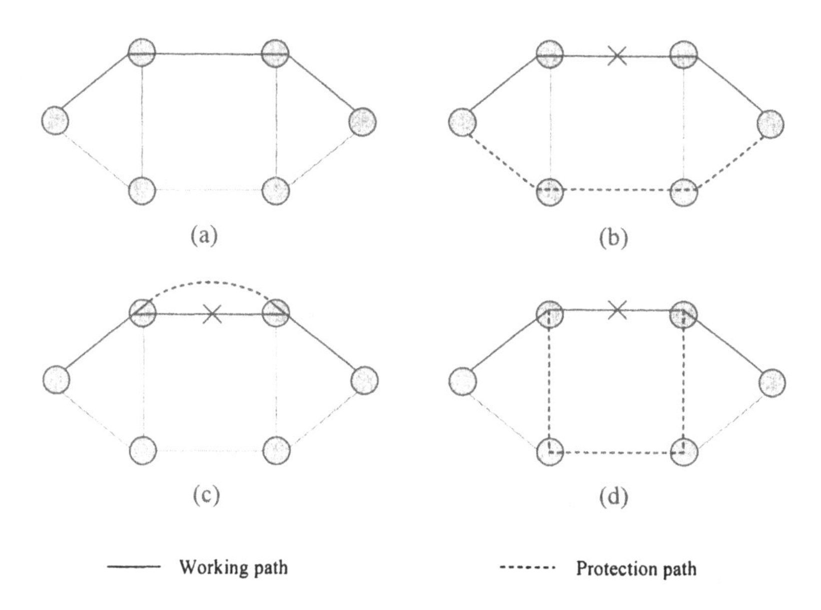

Figure 6.2 Path and link protection: (a) no protection; (b) path protection; (c) link (span) protection; (d) link (line) protection.

Link protection

In link protection, a backup path or protection path is reserved for each link of the primary path or working path during the establishment of a connection. In the event of a link failure, all the disrupted connections are rerouted around the failed link. The service recovery is handled at the end nodes of the failed link. The source and destination nodes are unconscious of the link failure and thus are not involved in the service recovery. In practice, there are two ways to implement link protection: span protection and line protection, which are illustrated in Figure 6.2(c) and Figure 6.2(d), respectively. Because the end nodes of the failed link can detect the link failure directly, link protection can provide very fast service recovery. It should be pointed out that in link protection the wavelength used for a backup path must be the same as that used for the primary path unless wavelength converters are used.

Path protection

In path protection, a backup path is reserved for the primary path on an end-to-end basis during the establishment of a connection, which is usually link-disjoint. In the event of a link failure, all the disrupted connections are rerouted on an end-to-end basis, as illustrated in Figure 6.2(b). The service recovery of a disrupted connection is handled at the source and destination nodes of that connection. Unlike link protection, in which the end nodes of a failed link can directly detect the link failure, path protection requires the end nodes of the failed link to inform the source and destination nodes of the link failure. This would result in a longer time in service recovery and require multiple nodes to cooperate together in an effective way, including not only the end nodes of the failed link and the source and destination nodes of the disrupted connection but also the intermediate nodes. In path protection, the wavelength used for a backup path does not have to be the same as that used for the primary path.

Link protection can be further classified into dedicated link protection and shared link protection.

Dedicated link protection

In dedicated link protection, for each link of the primary path a backup path and wavelength are reserved around that link during the establishment of a connection. The wavelength reserved on each link of the backup path is dedicated to that backup path.

Shared link protection

In shared link protection, for each link of the primary path a backup path and wavelength are reserved around that link during the establishment of a connection. However, the wavelength reserved on each link of the backup path may be shared among multiple backup paths.

Path protection can further be classified into dedicated path restoration and shared path restoration.

Dedicated path protection

In dedicated path protection, a backup path and wavelength are reserved for the primary path during the establishment of a connection. The wavelength reserved on each link of the backup path is dedicated to that backup path. Figure 6.3(a) illustrates dedicated path protection, in which a primary path is established on path 1-2-3-4 using wavelength 1 (λ_1) and the backup path is established on path 1-7-8-4 using wavelength 2 (λ_2). Wavelength 2 reserved on each link of the backup path is dedicated to the backup path.

Shared path protection

In shared path protection, a backup path and wavelength are reserved for the primary path during the establishment of a connection. The wavelength reserved on each link of the backup path may be shared among multiple backup paths. Figure 6.3(b) illustrates shared path protection, in which the backup path for one primary path is established on path 2-5-6-3 using wavelength 1 (λ_1) and the backup path for another primary path is established on path 7-5-6-8 also using wavelength 1 (λ_1). Obviously, wavelength 1 on link 5-6 is shared by the two backup paths.

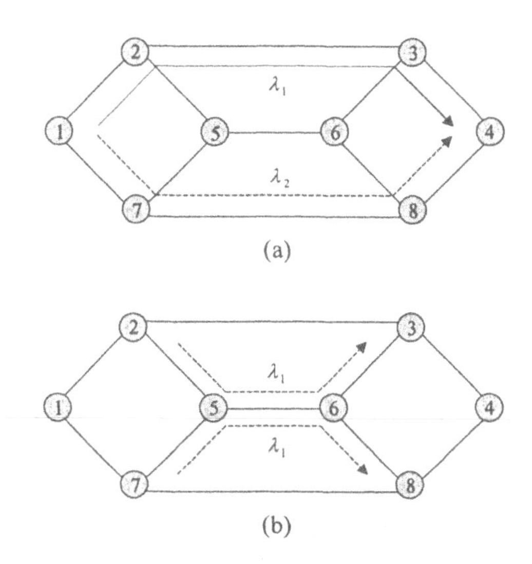

Figure 6.3 Path protection: (a) dedicated; (b) shared.

In addition, path protection can be classified into link-dependent protection and link-independent protection.

Link-dependent path protection

In link-dependent path protection, there is a different backup path to protect a primary path for each link of the primary path. In the event of a link failure, the backup path corresponding to the failed link is used to protect the primary path. Figure 6.4 illustrates link-dependent path protection. The backup paths for each link on the primary path are listed in Table 6.1.

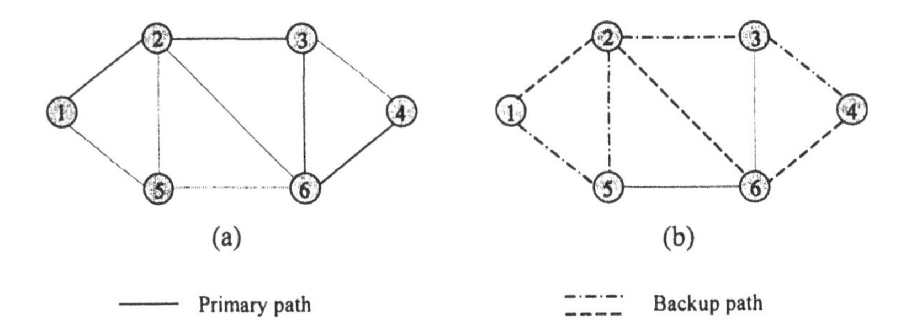

Figure 6.4 Link-dependent path protection: (a) primary path; (b) backup path.

Table 6.1

Link	Backup path
(1, 2)	1-5-2-3-4
(2, 3)	1-2-6-4
(3, 6)	1-2-6-4
(6, 4)	1-5-2-3-4

Link-independent path protection

In link-independent path protection, there is only a single backup path for a primary path. No matter which link of the primary path fails, the same backup path is used to protect the primary path. Figure 6.3(a) is also an example of link-independent path protection.

6.3.5 Link Restoration and Path Restoration

There are also two basic restoration paradigms for dynamic restoration: link restoration and path restoration.

Link restoration

In link restoration, the end nodes of the failed link dynamically search for a backup path around the link for each connection that traverses the failed link in the even of a link failure. If no backup path is discovered for a disrupted connection, the connection will be blocked.

Path restoration

In path restoration, the source and destination nodes of each connection that traverses the failed link dynamically search for a backup path on an end-to-end basis in the event of a link failure. If no backup is discovered for a disrupted connection, the connection will be blocked.

In general, path restoration is more efficient in service recovery than link restoration. This is because path restoration searches for a backup path on an end-to-end basis and the backup path can use a wavelength different from that used for a disrupted connection, whereas link restoration can only search for a backup path around the failed link on the same wavelength as that used for a disrupted connection. As a result, path restoration performs better in finding a backup path on an available wavelength. However, link restoration has a faster restoration speed than path restoration. In link restoration, the backup paths found tend to have fewer hops than those found in path restoration. On the other hand, there is no need for the end nodes of a failed link to inform the source and destination nodes of each disrupted connection. As a result, link restoration is faster in service recovery than path restoration.

A classification of the basic protection and restoration schemes for surviving single-link failures is summarized in Figure 6.5.

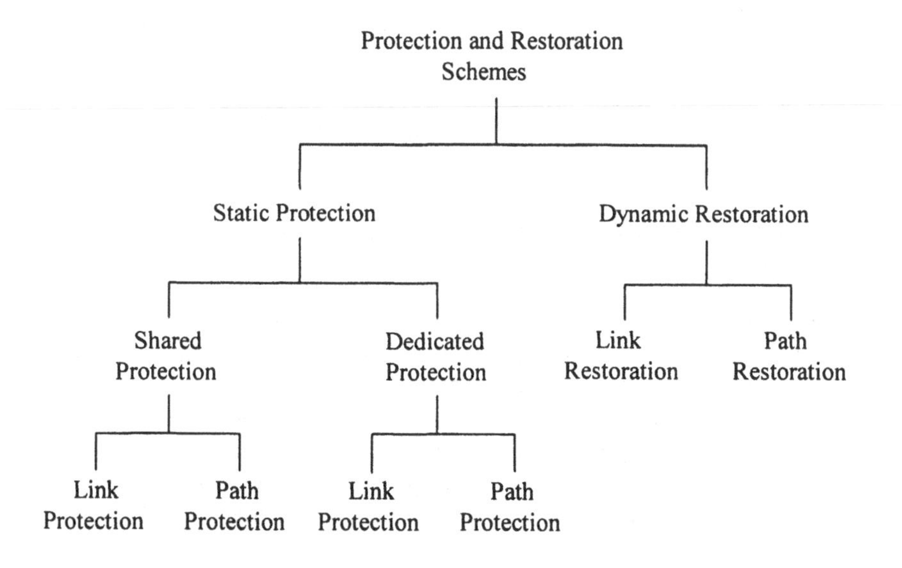

Figure 6.5 Basic protection and restoration schemes.

6.3.6 Segment Protection

In conventional path protection, a backup path is provisioned for a primary path on an end-to-end basis to handle any single-link failure on the primary path. This usually requires a couple of link-disjoint end-to-end paths with sufficient network resources between the source node and the destination node, one as the primary path and the other as the backup path. In practice, however, it is not always possible to find a link-disjoint backup path for each primary path in the network. In some situations, even though there are a couple of link-disjoint paths existing between the source and destination nodes, the primary path may be routed such that a link-disjoint backup path cannot be found in order to meet some other requirements. For example, the primary path may be routed along the minimum delay path to the destination node to reduce the propagation delay. On the other hand, link protection is inefficient in terms of resource utilization. For these reasons, segment protection recently has been proposed to address the problem [12–14].

In segment protection, a primary path is divided into a concatenation of shorter segments called primary segments. For each of the primary segments, a backup path is found independently, which is called a backup or protection segment. All the protection segments constitute a segmented protection path. Figure 6.6 illustrates the concept of segment protection for a primary path with nine nodes and eight links, where node 0 is the source node and node 8 is the destination node. The primary path is divided into three primary segments. Each of the primary segments has a protection segment and all three protection segments constitute the segmented protection path for the primary path, as shown in Figure 6.6(a). Two consecutive primary segments overlap at least by one link. When a link of a primary segment fails, the primary path is rerouted just around that primary segment using the corresponding protection segment. If only one primary segment covers the failed link, the protection segment corresponding to that primary segment is used to reroute the disrupted primary path, as shown in Figure 6.6(b). If two consecutive primary segments cover the failed link, any one of the two protection segments corresponding to the two primary segments can be used to recover the disrupted primary path, as shown in Figure 6.6(c). Note that conventional end-to-end path protection is actually a special case of segment protection, in which the number of primary segments is equal to one. It was shown in [13] that segment protection has the following advantages over conventional end-to-end path protection:

- Faster in service recovery
- More efficient in resource utilization
- More flexible in providing dependable connections

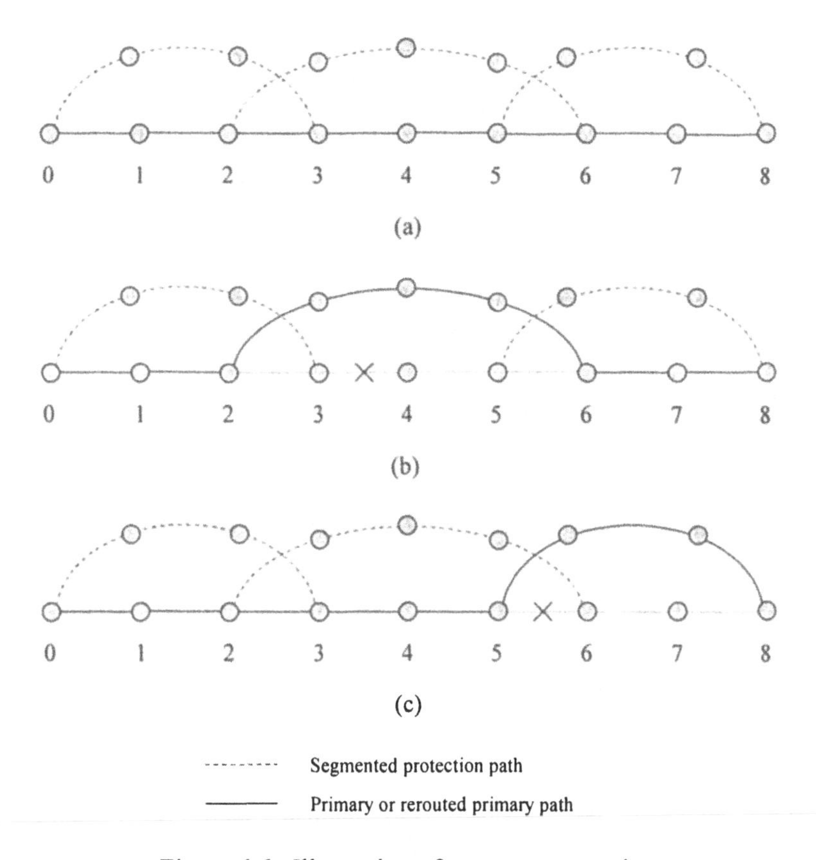

Figure 6.6 Illustration of segment protection.

6.3.7 Considerations in Survivable Network Design

The traffic demand may have a different impact on the design of survivable networks. For wavelength-routed WDM networks, there are two basic types of traffic demand: static traffic and dynamic traffic. For static traffic demand, a set of connection requests is known a priori. The survivable network design problem is to establish a set of primary lightpaths to accommodate the traffic demand and meanwhile to provide a set of backup paths to protect against network failures. The objective is to minimize the spare network resources used to establish the backup paths, or in other words, to establish as many primary paths with protection capability as possible for a given amount of network resources. For dynamic traffic demand, connection requests arrive to the network randomly. To provide service protection against network failures, the network must dynamically establish a primary path and a protection path for each connection request, which is referred to as the survivable routing problem. If either a primary path or a protection path cannot be established for a connection request, the request will be blocked. Accordingly, the objective of survivable routing is to minimize the connection blocking probability in the network.

On the other hand, network control can be either centralized or distributed, which may also have a different impact on the design of survivable networks. Under centralized control, there is a centralized controller in the network, which is responsible for the recovery of disrupted network services in the event of network failures. Under distributed control, there is no centralized controller in the network. The service recovery in the event of network failures is handled at each network node.

6.4 Survivable Network Design for Static Traffic

In this section, we discuss the survivable network design problem for static traffic.

6.4.1 Survivable Network Design Problem

For static traffic demand, a set of connection requests is known a priori. Given the static traffic demand as well as the network physical topology, the survivable network design problem is to establish a set of primary lightpaths to accommodate the traffic demand and meanwhile provide a backup path for each primary path to protect network services against network failures. In survivable network design, the objective is to minimize the spare network resources used to provide the backup paths or to minimize the number of wavelengths used for both primary paths and backup paths, while maintaining a high level of protection capability. For this purpose, various protection schemes can be used, such as dedicated protection, shared protection, link protection, and path protection. This network design problem can be generally formulated as an integer linear programming (ILP) problem. For small networks of a few tens of nodes, ILP can be used to find an optimal solution to the problem. However, for large networks of a few hundreds of nodes, an ILP problem becomes computationally intractable. Heuristics are usually used to obtain approximate solutions. In the next section, we will present two ILP formulations proposed in [5], which are based on dedicated path protection and shared path protection, respectively.

6.4.2 ILP Formulations

In the following formulations, the network physical topology and the traffic demand in terms of a set of connection requests are given. A set of alternate routes is predetermined for each pair of network nodes. These alternate routes are used for both primary paths and backup paths. A set of parameters is also given. The notations used in the formulations are defined as follows.

Parameters

- N: the number of nodes in the network
- M: the number of links in the network
- W: the number of wavelengths on each fiber link
- λ_k: the number of connection requests between node pair k
- C^k: the number of alternate routes between node pair k
- C_m: the maximum number of alternate routes between any node pair
- $\boldsymbol{R}^k$: a set of alternate routes between node pair k
- $\boldsymbol{R}_i^k$: a set of alternate routes used for backup paths between node pair k in the event of a failure on link i

Variables

- w_i^p : the number of wavelengths used for primary paths on link i
- w_i^b : the number of wavelengths used for backup paths on link i
- $p_w^{k,r}$: takes a value of 1 if the rth route between node pair k is used as a primary path on wavelength w; otherwise 0
- $b_w^{k,r}$: takes a value of 1 if the rth route between node pair k is used as a backup path on wavelength w; otherwise 0
- $\alpha_{w,i}^{k,r}$: takes a value of 1 if wavelength w is used by backup path r between node pair k when link i fails; otherwise 0
- β_w^i : takes a value of 1 if wavelength w is used by backup path r that traverses link i; otherwise 0

Formulation 6.1

Objective
$$Min[\sum_{i=1}^{M}(w_i^p + w_i^b)] \tag{6-1}$$

subject to

$$w_i^p + w_i^b \le W \qquad 1 \le i \le M \tag{6-2}$$

$$\lambda_k = \sum_{r=1}^{C^k} \sum_{w=1}^{W} p_w^{k,r} \qquad 1 \le k \le N(N-1) \tag{6-3}$$

$$w_i^p = \sum_{k=1}^{N(N-1)} \sum_{r \in R^k, i \in r} \sum_{w=1}^{W} p_w^{k,r} \qquad 1 \le i \le M \tag{6-4}$$

$$w_i^b = \sum_{k=1}^{N(N-1)} \sum_{r \in R^k, i \in r} \sum_{w=1}^{W} b_w^{k,r} \qquad 1 \le i \le M \tag{6-5}$$

$$\sum_{k=1}^{N(N-1)} \sum_{i \in r, r \in R^k} (p_w^{k,r} + b_w^{k,r}) \le 1 \tag{6-6}$$

$$1 \le w \le W, 1 \le i \le M$$

$$\sum_{r \in R^k, i \in r} \sum_{w=1}^{W} p_w^{k,r} = \sum_{r \in R_i^k} \sum_{w=1}^{W} b_w^{k,r} \tag{6-7}$$

$$1 \le i \le M, 1 \le k \le N(N-1)$$

This formulation is based on dedicated path protection. Objective (6-1) is to minimize the number of wavelengths used for both primary paths and backup paths. Constraint (6-2) specifies that the total number of primary paths and backup paths on each link is bounded to the number of wavelengths available on that link. Constraint (6-3) specifies that the number of connection requests between node pair k can be accommodated with the number of primary paths established. Constraint (6-4) defines the number of wavelengths used for primary paths on each link. Constraint (6-5) defines the number of wavelengths used for backup paths on each link. Constraint (6-6) specifies that only one primary path or backup path can use wavelength w on link i, i.e., the wavelength-continuity constraint. Constraint (6-7) specifies that in the event of a link failure, the number of connection requests between each node pair can still be accommodated.

Formulation 6.2

Objective

$$Min[\sum_{i=1}^{M} (w_i^p + w_i^b)] \tag{6-8}$$

subject to

$$w_i^p + w_i^b \leq W \qquad 1 \leq i \leq M \tag{6-9}$$

$$\lambda_k = \sum_{r=1}^{C^k} \sum_{w=1}^{W} p_w^{k,r} \qquad 1 \leq k \leq N(N-1) \tag{6-10}$$

$$w_i^p = \sum_{k=1}^{N(N-1)} \sum_{i \in r, r \in R^k} \sum_{w=1}^{W} p_w^{k,r} \qquad 1 \leq i \leq M \tag{6-11}$$

$$w_i^b = \sum_{w=1}^{W} \beta_w^i \qquad 1 \leq i \leq M \tag{6-12}$$

$$\beta_w^j \leq \sum_{k=1}^{N(N-1)} \sum_{i=1}^{M} \sum_{j \in r, r \in R_i^k} \alpha_{w,i}^{k,r} \tag{6-13}$$

$$1 \leq j \leq M, 1 \leq w \leq W$$

$$N \times (N-1) \times M \times C_m \times \beta_w^j \geq \sum_{k=1}^{N(N-1)} \sum_{i=1}^{M} \sum_{j \in r, r \in R_i^k} \delta_{w,i}^{k,r} \tag{6-14}$$

$$1 \leq j \leq M, 1 \leq w \leq W$$

$$[\sum_{k=1}^{N(N-1)} \sum_{i \in r, r \in R^k}^{W} p_w^{k,r}] + \beta_w^i \leq 1 \tag{6-15}$$

$$1 \leq i \leq M, 1 \leq w \leq W$$

$$\sum_{r \in R_i^k} \sum_{w=1}^{W} \alpha_{w,i}^{k,r} = \sum_{i \in r, r \in R^k} \sum_{w=1}^{W} p_w^{k,r} \tag{6-16}$$

$$1 \leq i \leq M, 1 \leq k \leq N(N-1)$$

This formulation is based on shared path protection. Objective (6-8) is also to minimize the number of wavelengths used for both primary paths and backup paths. Constraint (6-9) specifies that the total number of primary paths and backup paths on each link is bounded to the number of wavelengths available on the link. Constraint (6-10) specifies that the number of connection requests between node pair k can be accommodated with the number of primary paths established. Constraint (6-11) defines the number of wavelengths used for primary paths on each link. Constraint (6-12) defines the number of wavelengths used for backup paths on each link. Constraints (6-13) and (6-14) indicate whether wavelength w is used by some backup path on link j. Constraint (6-15) is the wavelength-continuity constraint, which specifies that only one primary or backup path can use wavelength w on link i. Constraint (6-16) defines the number of rerouted primary paths between node pair k when link i fails.

The above ILP formulations are only two typical examples of ILP formulations for survivable network design, which have been shown to be NP complete [5] and thus are only tractable for small networks. For large networks, it is not practical to find an optimal solution to such ILP formulations. For this reason, heuristics are often used to obtain approximate solutions. In addition to the above examples, many different ILP formulations have been proposed in the literature for survivable network design [15–21]. These ILP formulations address different network scenarios (e.g., single-fiber networks, multi-fiber networks, wavelength-selective networks, or wavelength-convertible networks) and different types of network failures (e.g., link failures or node failures) with different protection schemes (e.g., dedicated protection or shared protection).

6.5 Survivable Routing for Dynamic Traffic

For dynamic traffic, each connection request arrives to the network randomly on a one-by-one basis. To provide service protection against network failures, a backup path must be established simultaneously at the time of establishing a primary path for each connection request. The concept of survivable routing is to simultaneously route both a primary path and a backup path to accommodate a connection request. If either a primary path or a backup path cannot be established for a connection request, the request will be blocked. The objective of survivable routing is to successfully route both the primary path and the backup path with the minimum consumption of network resources. For this purpose, survivable routing can be based on various protection schemes (e.g., dedicated protection and shared protection) and routing strategies (e.g., static strategy, dynamic strategy, separate strategy, and joint strategy) and can be performed under either centralized control or distributed control. In this section, we will present some typical routing algorithms and strategies that have already been proposed in the literature for survivable routing.

6.5.1 Dedicated and Shared Survivable Routing

Survivable routing can be based on either dedicated protection or shared protection. In survivable routing, a link or node disjoint backup path must be established at the time of establishing a primary path for each connection request. In dedicated protection the backup path is dedicated to the primary path, whereas in shared protection the backup path can be shared among multiple backup paths. In general, routing for dedicated protection is relatively simpler to perform. A routing algorithm can simply compute a route with the minimum cost for a primary path first, remove the links of the primary path, and then compute a route with the minimum cost for a backup path. For shared protection, routing is more complicated. To achieve a trade-off between service protection and network performance, some requirements must be taken into account in the design of a survivable routing algorithm. The major requirements for survivable routing are described as follows.

- A primary path must not share the same risk with its backup path, or in other words, a primary path and its backup path must not have links belonging to the same shared risk link group (SRLG) [22], which is referred to as the risk disjoint constraint (or the SRLG constraint). Otherwise, a single failure may cause the disruption of both paths and thus result in unsuccessful service recovery. This requirement applies to both dedicated protection and shared protection.
- Any two primary paths sharing the same risk or link belonging to the same SRLG must not have their backup paths share the same spare resources. Otherwise, a single failure may cause the disruption of two or more primary paths and result in no guarantee of protection for all the disrupted primary paths.
- The spare resources that have already been used by a backup path should be shared as much as possible with other backup paths in order to increase the resource utilization in the network. Such shared resources should be given higher preferences in routing a backup path.

Note that an SRLG is defined as an administrative group associated with some fiber links that are potentially vulnerable to a single failure. For example, all fibers in the same conduit belong to the same SRLG because a conduit cut will result in the failure of all the fibers in the conduit.

6.5.2 Cost Function

In survivable routing, a routing algorithm usually computes a route with the minimum cost for the primary path and the backup path based on a cost function that measures the cost of a path. The cost function plays an important role in survivable routing. Usually, the cost of a path depends on the cost of each link that the path traverses. The definitions of the link cost can be classified into three basic types:

- Topology-based cost
- Resource-based cost
- Integrated cost

The topology-based link cost is usually defined in terms of the hop count, link distance, etc. For example, if the hop count is used, the cost of a link can be defined as one. If the link distance is used, the cost of a link can be defined as the link distance. The resource-based link cost is usually defined in terms of the available wavelengths or other network resources such as wavelength converters. For example, if there is no wavelength available on a link, the link cost may be defined as infinity. The integrated link cost combines both network topology information and resource usage information into the cost in a particular way. In most cases, the hop count and the available wavelengths are combined in some weighted form to define the link cost.

> ***Example 6.1***: An integrated link cost combining both the hop count and wavelength usage information is defined as follows [23].
>
> $$C_l = \alpha \times f(n_1, n_2) + \beta \times 1$$
>
> where C_l presents the cost of link l; n_l is the total number of used wavelengths and n_2 is the total number of wavelengths on each fiber link; α and β are the weights on the wavelength usage and the hop count, respectively, which are used to control the relative significance of the hop count or the wavelength usage information; The number "1" represents the hop count cost; $f(n_l, n_2)$ is a function that controls the effect of n_l and n_2 in the cost function, which can be any appropriate linear or nonlinear function.

The path cost can be defined on the basis of the link cost. For the primary path, the path cost is simply the total cost of all the links that the path traverses. For the backup path, however, the path cost depends on the protection scheme used. For dedicated protection, the backup path cost is similar to the primary path cost, which is defined as the total cost of all the links that the backup path traverses. For shared protection, the backup path cost is not only related to the link cost but more related to the resource sharing. If a backup path shares a wavelength on a particular link with other backup paths, the contribution of the cost of this particular link to the backup path cost should be significantly reduced, resulting in the link being more likely to be shared by other backup paths.

> ***Example 6.2***: A primary path cost C_p and a backup path cost C_b are defined as follows
>
> $$C_p = \sum_{l \in p} C_l$$
>
> $$C_b = \sum_{l \in b} g(l) C_l$$
>
> where l represents a link; p and b represent the primary path and the backup path, respectively; $g(l)$ is a weight function related to the resource sharing. If link l is not shared, $g(l)=1$. Otherwise, $0 \leq g(l) < 1$.

6.5.3 Static and Dynamic Survivable Routing

In survivable routing, a routing algorithm usually uses a static strategy for establishing backup paths and most survivable routing algorithms are actually based on the static strategy. With the static strategy, once a backup path is established, both the route and wavelength used for the backup path are not allowed to change, which is simple to implement but not flexible. To provide more flexibility, a dynamic strategy was proposed in [24], which allows the route and wavelength of a backup path to change dynamically after the backup path is established. This implies that the route and wavelength already used for a particular backup path can still be available to a new primary path and any other backup path as long as the backup path can be reestablished on a different wavelength and/or a different route. With this strategy, all primary paths are still protected and no service disruption is caused during the reestablishment of a backup path.

Survivable routing algorithm based on a static strategy

A survivable routing algorithm based on a static strategy is described as follows [24].

- For each pair of source and destination nodes, use *Dijkstra*'s algorithm to compute the shortest path (in terms of hops) and use this shortest path as the route for the primary path.
- Remove all the links along the shortest path, then use *Dijkstra*'s algorithm again to compute the shortest path, and use this shortest path as the route for the backup path.

- For a connection request, assign an available wavelength to the predetermined primary path. If no available wavelength can be assigned, the connection request will be blocked.
- After an available wavelength is assigned to the primary path, assign an available wavelength to the predetermined backup path. If no available wavelength can be assigned, the connection request will also be blocked. Otherwise, the connection request will be successfully accommodated.

Survivable routing algorithm based on a dynamic strategy

A survivable routing algorithm based on a dynamic strategy is described as follows [24].

- For each pair of source and destination nodes, predetermine the primary path and the backup path in the same manner as that used with the static strategy.
- For a connection request, assign an available wavelength to the predetermined primary path. If no available wavelength can be assigned, the connection request will be blocked. Note that the wavelengths already assigned to all established backup paths are still considered available to the primary path as long as the backup paths can still be established.
- After an available wavelength is assigned to the primary path, assign an available wavelength to the predetermined backup path. If no available wavelength can be assigned, the connection request will also be blocked. Otherwise, the connection request will be successfully accommodated.

To better understand the two strategies for survivable routing, let us examine the following example.

Example 6.3: Consider a network with five nodes and six links and two wavelengths on each fiber link, as shown in Figure 6.7. The network can be represented by two graphs, each for one wavelength. Suppose that there are three connection requests to the network with request 1 arriving first, followed by request 2 and request 3. The primary path and backup path for the requests are represented by P_i and B_i $(1 \leq i \leq 3)$, respectively. Figure 6.7(a) and Figure 6.7(b) illustrate the routing and wavelength assignment with the static strategy and the dynamic strategy, respectively, for the three connection requests. With the static strategy, P_1 and B_1 are established on path 2-1 and path 2-3-1, respectively, with wavelength 1. P_2 and B_2 are established on path 4-2 and path 4-5-3-2, respectively, also with wavelength 1. P_3 and B_3 are established on path 4-5 and path 4-2-3-5, respectively, with wavelength 2. With the dynamic strategy, P_1 and B_1 are still established on path 2-1 and path 2-3-1, respectively, with wavelength 1. P_2 and B_2 are still originally established on path 4-2 and path 4-5-3-2, respectively, with wavelength 1. For P_3 and B_3, note that wavelength 1 on path 4-5 is still considered available because B_2 can be reestablished on path 4-5-3-2 with wavelength 2. Accordingly, P_3 can be established on path 4-5 with wavelength 1 while B_3 can be established on path 4-2-3-5 with wavelength 2. Although both strategies can accommodate the three connection requests, the wavelengths assigned for the connections are different. This may affect the number of connection requests that can be accommodated by the two strategies.

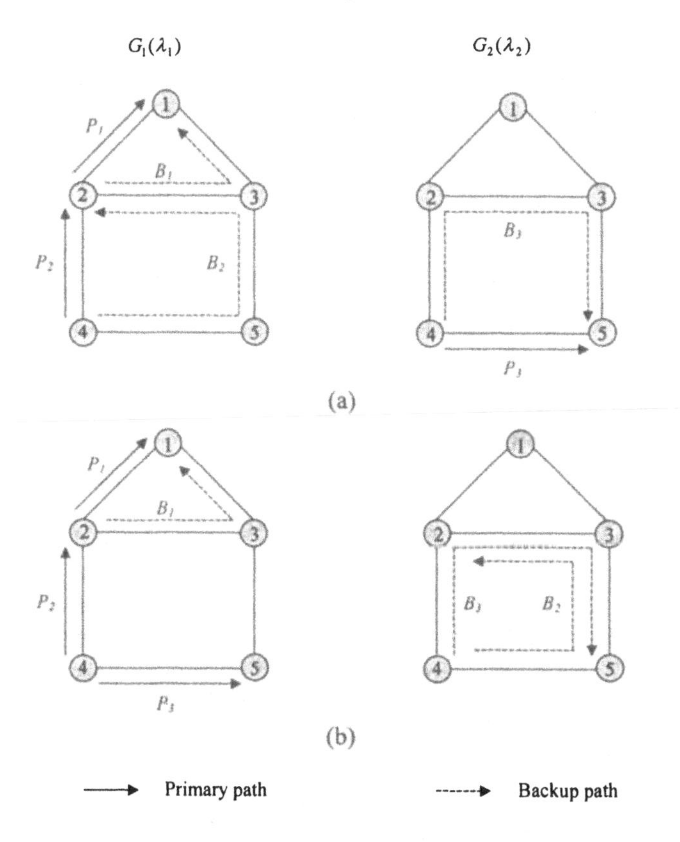

Figure 6.7 Survivable routing with: (a) static strategy; (b) dynamic strategy.

6.5.4 Separate and Joint Survivable routing

A survivable routing algorithm can also be based on either a separate routing strategy or a joint routing strategy [23]. With the separate strategy, a routing algorithm computes a primary path first and then computes a backup path for each connection request. A shortest-path routing algorithm is usually used to compute a route with the minimum cost for the primary path and the backup path. This strategy does not consider any optimization at all. Instead, it just attempts to find a route with the minimum cost. As a result, the primary and backup paths are computed separately with the minimum costs but the total cost of both paths may not be minimized. For dynamic traffic, it is not practical to perform global optimization in survivable routing. However, it is possible to perform local optimization in routing the primary path and the backup path for each connection request. If the total cost of both paths can be minimized for each connection request, the resource utilization of the network can still be optimized to a certain extent and better network performance can be achieved. For this reason, a joint routing strategy can be used in a survivable routing algorithm to jointly compute a route for the primary path and the backup path with the objective of minimizing the total cost of both paths.

Survivable routing algorithm based on a separate strategy

A survivable routing algorithm based on a separate routing strategy is described as follows.

- For each connection request, compute a route with the minimum cost and select this path as the primary path.
- For the primary path, compute a link-disjoint route with the minimum cost and select this route as the backup path.

Example 6.4: Consider a network that has six nodes and nine links with several wavelengths on each fiber link, as shown in Figure 6.8. The label on each link represents the cost for using the link. Suppose that there is a connection request for a lightpath from node *a* to node *d*. The minimum cost route from node *a* to node *d* is *a-b-e-d*, which has a cost of 3. With the survivable routing algorithm based on the separate strategy, this route is selected as the primary path. For the primary path, a link disjoint route with the minimum cost is *a-f-e-c-d*, which has a cost of 9. Accordingly, this route is selected as the protection path, as shown in Figure 6.8(a). As a result, the total cost of both paths is 12.

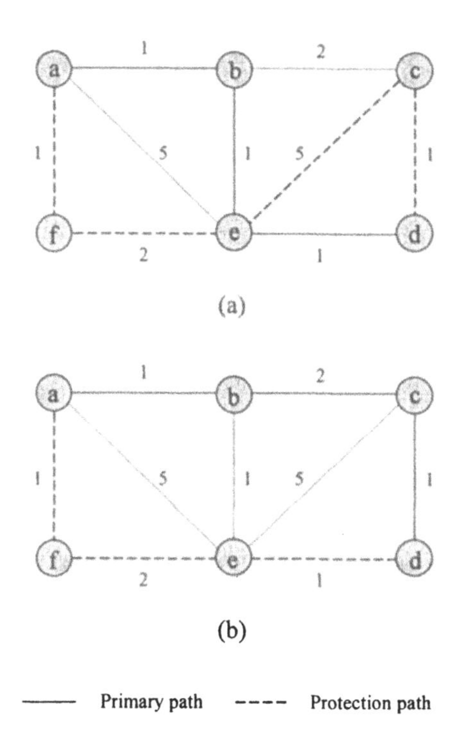

Figure 6.8 Survivable routing with: (a) separate strategy; (b) joint strategy.

Survivable routing algorithm based on a joint strategy

A survivable routing algorithm based on a joint routing strategy is described as follows.

- For each connection request, compute a set of alternate routes with the minimum cost
- For each of the alternate routes, compute a link-disjoint route with the minimum cost
- Calculate the total cost of each route and its associated link-disjoint route
- Find the route and its associated link-disjoint route with the minimum total cost
- Select the route as the primary route and its link-disjoint route as the backup path.

Example 6.5: Consider again the network in Figure 6.8. Also, suppose that there is a connection request for a lightpath from node *a* to node *d*. With the survivable routing algorithm based on the joint strategy, a set of alternate routes with the minimum cost is first computed and for each of the alternate routes, a link-disjoint route with the minimum cost is then computed, which are listed in Table 6.2.

Table 6.2

Route	Disjoint route	Total cost
a-b-e-d	a-f-e-c-d	3+9
a-b-c-d	a-f-e-d	4+4
a-f-e-d	a-b-c-d	4+4
a-e-d	a-b-c-d	6+4

Obviously, *a-b-c-d* and *a-f-e-d* are the routes with a minimum total cost of 8. Accordingly, one of them is selected as the primary path and the other is selected as the protection path, as shown in Figure 6.8(b). The primary path and protection path selected with the joint strategy have a smaller total cost than those selected with the separate strategy.

6.5.5 Centralized and Distributed Survivable Routing

To address different network scenarios, survivable routing can be performed under either centralized control or distributed control. In a centralized control scenario, there is a centralized controller in the network, which is responsible for the lightpath control, including the provisioning of both primary path and backup path for each connection request on behalf of all network nodes. Because the centralized controller maintains global network state information and has full knowledge of the resource usage and sharing on each link, survivable routing under centralized control is relatively simpler to implement compared with distributed control. The centralized controller only needs to perform a routing algorithm to compute a primary path as well as a backup path for each connection request. In the literature, a variety of algorithms are proposed for survivable routing under centralized control [10][25–30]. However, these routing algorithms are not scalable and thus are only suitable for small networks. For large networks, distributed survivable routing is preferred.

In a distributed control scenario, there is no centralized controller in the network. Survivable routing is performed at each network node. Each node must perform a survivable routing algorithm to compute a primary path and a backup path for each connection request. If a node does not have global network state information, it is unable to decide the shareablility of a link for any particular backup path and thus unable to support shared protection in the network. As we have already seen, shared protection provides a good trade-off between service recovery time and network resource utilization and is therefore highly preferred. To support shared protection under distributed control, it is necessary to distribute global network state information throughout the network. For this purpose, the network must use a link state advertisement protocol to distribute the resource usage and sharing information on each link in an effective manner. An example of link state advertisement protocols can be found in [31] and [32]. In some cases, additional information may also be needed for making a routing decision, which can be distributed with some signaling protocols [33]. On the basis of such link state and additional information, a survivable routing algorithm can make a more intelligent routing decision to provision a shared backup path for a particular primary path, which can thus achieve better resource sharing in the network. Therefore, distributed survivable routing is more complex to implement. A good example of distributed survivable routing can be found in [10], in which both routing algorithms and signaling protocols for survivable routing are described.

Survivable routing has been studied extensively for wavelength-routed WDM networks. A variety of survivable routing algorithms have already been proposed in the literature, which are based on different cost functions and protection schemes to achieve different trade-offs between service recovery time and network resource utilization. The readers are referred to [10] and [25–30] for a detailed description of these algorithms.

6.6 Dynamic Restoration

We have already discussed the survivable network design for static and dynamic traffic based on static protection. In this section, we discuss dynamic restoration and present several typical dynamic restoration mechanisms for handling single-link failures.

6.6.1 Problems in Dynamic Restoration

Unlike static protection, dynamic restoration does not reserve any network resources. Instead, the network must search dynamically for spare network resources available in the network to recover the disrupted network services after network failures occur. As a result, dynamic restoration is more efficient in resource utilization than static protection but is slower in service recovery. Moreover, the service recovery with dynamic restoration may not be guaranteed because of insufficient spare network resources in the network. For this reason, fast discovery and provisioning of spare network resources for service recovery in the event of network failures have been an important problem with dynamic restoration.

As mentioned in Section 6.3.5, dynamic restoration can be classified into link restoration and path restoration. The main difference lies in the way a disrupted connection is rerouted or in the nodes that are responsible for the service recovery. In link restoration, the end nodes of a failed link are responsible for the discovery and provisioning of a backup path around the link for each disrupted connection. In path restoration, the source and destination nodes of each disrupted connection are responsible for the discovery and provisioning of a backup path on an end-to-end basis. Because of this difference, link restoration and path restoration may result in different network performance. In general, link restoration provides faster service recovery whereas path restoration has better recovery efficiency [6].

In both link restoration and path restoration, the discovery and provisioning of a backup path involve the following four problems:

- Routing
- Wavelength assignment
- Wavelength reservation
- Restoration initiation

The first three problems are similar to those involved in the establishment of a primary path. In addition to these three problems, an important problem that is unique to dynamic restoration is the restoration initiation problem. In dynamic restoration, not only the source node of each disrupted connection but also the destination node and both the end nodes of a failed link can initiate a restoration process. In some situations, the intermediate nodes of the primary path of a disrupted connection may also initiate a restoration process. For this reason, the restoration initiation problem may have a big impact on the network performance in terms of connection restoration time and connection restoration probability. In the next sections, we will present several dynamic restoration mechanisms already proposed for dynamic restoration, including source-initiated restoration, destination-initiated restoration, bi-initiation-based restoration, and multi-initiation-based restoration.

6.6.2 Source-Initiated Restoration

Source-initiated restoration is the most commonly used restoration mechanism for dynamic restoration, which applies to both link restoration and path restoration. In link restoration, the source-side end node of a failed link initiates a connection restoration process on the detection of the link failure. In path restoration, the source node of each disrupted connection initiates a connection restoration process on being notified of a link failure. In a restoration process, a backup path is established for a disrupted connection in a way similar to the establishment of a primary path by employing a dynamic restoration protocol. The dynamic restoration protocol can combine different routing algorithms [34] and wavelength reservation protocols [35] to achieve fast service recovery in different network scenarios.

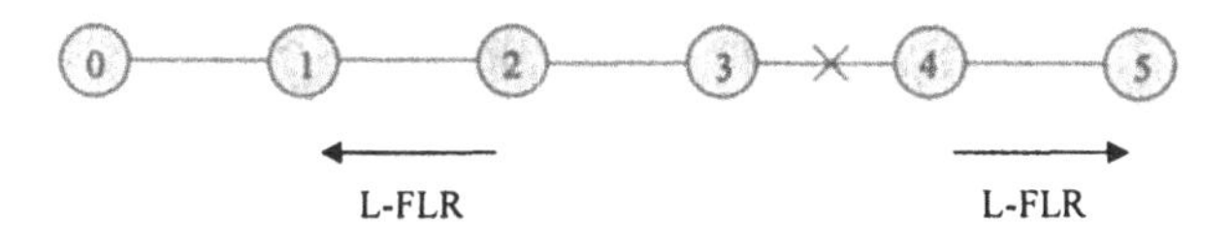

Figure 6.9 Illustration of link failure notification.

Example 6.6: A source-initiated path restoration protocol based on source routing and backward reservation is briefly described as follows.

- Once the end nodes of a failed link detect the link failure, both nodes send a link failure (*L-FLR*) packet to the source and destination nodes of each disrupted connection, as shown in Figure 6.9. For each disrupted connection, the *L-FLR* packet disconfigures the optical switch and releases the wavelength reserved for the connection at each intermediate node.

- Once the source node of a disrupted connection receives the *L-FLR* packet, it initiates a restoration process for the disrupted connection.

- In the restoration process, the source node first performs a routing algorithm to decide on a new route for the disrupted connection and then sends a request (*S-REQ*) packet to the destination node along the decided route, as shown in Figure 6.10. At each intermediate node, the *S-REQ* packet collects wavelength usage information on each link of the decided route.

- When the destination node receives the *S-REQ* packet, it will select an available wavelength based on the collected information and then send a reservation (*D-RSV*) packet to the source node along the reserve route. At each intermediate node, the *D-RSV* packet attempts to reserve the selected wavelength and simultaneously configure the optical switch. If the wavelength can be reserved, the *D-RSV* packet will be forwarded to the next hop. Otherwise, a negative acknowledgment (*I-NAK*) packet will be sent to the destination node. The *I-NAK* packet will disconfigure the optical switch and release the wavelength already reserved by the *D-RSV* packet at each intermediate node.

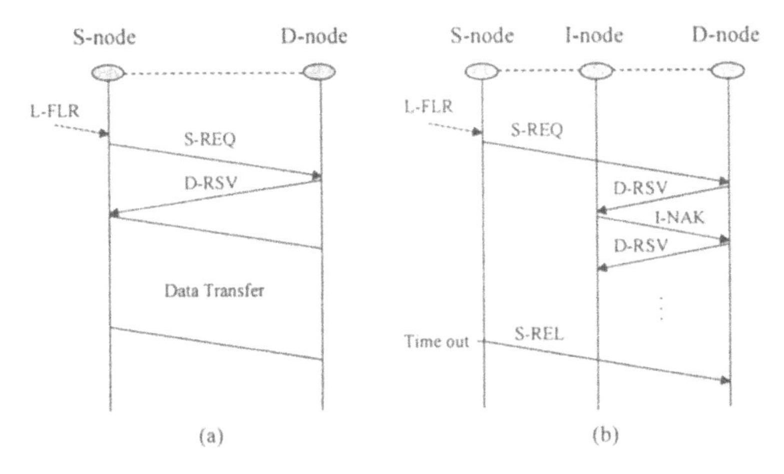

Figure 6.10 Source-initiated path restoration: (a) successful; (b) unsuccessful.

- When the destination node receives the *I-NAK* packet, it will attempt to select another available wavelength for reservation. If there is no wavelength available, the destination node will send a negative acknowledgement (*D-N*AK) packet to the source node.
- When the source node receives the *D-NAK* packet, it will terminate the restoration process and block the disrupted connection.
- If the *D-RSV* packet reaches the source node, it implies that a backup path has been established successfully for the disrupted connection.

- To control the restoration time, the source node uses a timer and sets the timer to the maximum allowed restoration time at the beginning of the restoration process. If the timer times out without receiving a *D-RSV* packet, the source node will block the disrupted connection and send a release (*S-REL*) packet to the destination node to terminate the restoration process.

6.6.3 Destination-Initiated Restoration

In source-initiated restoration, it takes a two-way delay to discover a backup path. A forward control packet must first be sent to the destination node, followed by a backward packet sent back to the source node. This may not be the most efficient way in both path restoration and link restoration. In path restoration, for example, the destination node of a disrupted connection may have a record of information about the disrupted connection, such as the connection identifier, the source address, and the wavelength used for the connection, at the time of the connection establishment. Accordingly, a connection restoration process does not have to be initiated by the source node. The destination node is also able to initiate a connection restoration process. This makes it possible to take a one-way delay to establish a backup path and thus reduce the connection restoration time significantly.

Following this observation, a destination-initiated restoration mechanism has been proposed in [36], which applies to both path restoration and link restoration. In path restoration, the destination node of each disrupted connection initiates a connection restoration process on being notified of a link failure. In link restoration, the destination-side end node of a failed link initiates a connection restoration process on the detection of a link failure. Based on this restoration mechanism, a destination-initiated path restoration protocol and a destination-initiated link restoration protocol were proposed in [36] and [37], respectively. It has been shown that the destination-initiated restoration can significantly reduce the connection restoration time as compared with the source-initiated restoration.

> ***Example** 6.7*: A destination-initiated path restoration protocol based on destination routing and backward reservation is briefly described as follows.

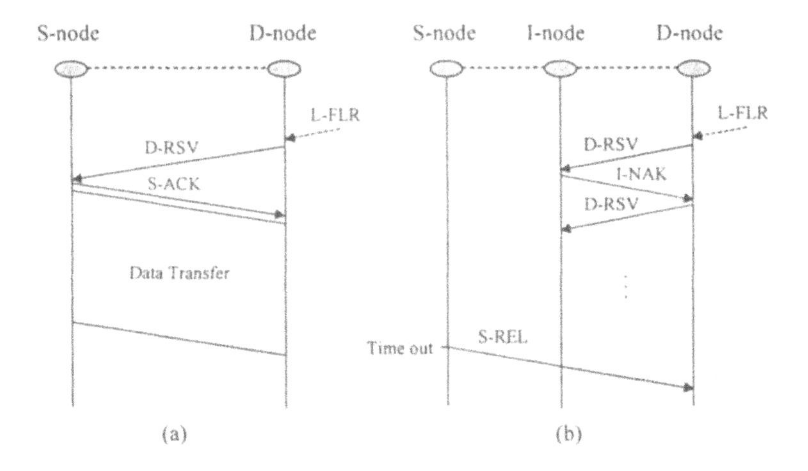

Figure 6.11 Destination-initiated path restoration: (a) successful; (b) unsuccessful.

- Once the destination node of a disrupted connection receives the *L-FLR* packet, it initiates a connection restoration process. In the restoration process, the destination node first performs a routing algorithm to decide on a new route and select an available wavelength for the disrupted connection, and then sends a reservation (*D-RSV*) packet to the source node along the reverse route, as illustrated in Figure 6.11.
- At each intermediate node, the *D-RSV* packet attempts to reserve the selected wavelength and simultaneously configure the optical switch. If the selected wavelength can be reserved, the *D-RSV* packet will be forwarded. Otherwise, a negative acknowledgment (*I-NAK*) packet will be sent back to the destination node. The *I-NAK* packet will disconfigure the local optical switch and release the wavelength already reserved by the *D-REQ* packet at each intermediate node.
- When the destination node receives the *I-NAK* packet, it will attempt to select another available wavelength on the same route or on another route. If there is no wavelength available, the destination node will send a negative acknowledgment (*D-N*AK) packet to the source node.
- Once the source node receives the *D-NAK* packet, it will block the disrupted connection.

- If the *D-REQ* packet reaches the source node, it implies that a backup path has been established successfully for the disrupted connection. In response, the source node will send an *S-ACK* packet to the destination node to terminate the restoration process. The source node can also use a timer to control the restoration time in a manner similar to that described in Section 6.6.2.

6.6.4 Bi-initiation-Based Restoration

The destination-initiated restoration mechanism can significantly reduce the connection restoration time compared with the source-initiated restoration mechanism. However, there still exist some particular cases in which the destination-initiated mechanism may result in a larger restoration time than the source-initiated mechanism. To further reduce the connection restoration time, a bi-initiation-based restoration mechanism was proposed in [38] that allows both the source node and destination node of each disrupted connection to respectively initiate a connection restoration process on being notified of a link failure. It has been shown that the bi-initiation-based restoration mechanism can significantly reduce the connection restoration time compared with the source-initiated and destination-initiated restoration mechanisms.

Example 6.8: A bi-initiation-based path restoration protocol [38] is briefly described as follows.

- Once the source node of a disrupted connection receives an *L-FLR* packet, it initiates a connection restoration process. In the restoration process, the source node first performs a routing algorithm to decide on a new route for the disrupted connection and then sends a request (*S-REQ*) packet to the destination node along the decided route.
- At each intermediate node, the *S-REQ* packet collects the wavelength usage information on each link of the decided route. If the destination node receives the *S-REQ* packet after it receives an *L-FLR* and sends out a reservation (*D-RSV*) packet, it will simply store the *S-REQ* packet and then wait for the next control packet. Otherwise, it will select an available wavelength based on the collected information, send an *S-RSV* packet back to the source node along the reverse route, and then store the *S-REQ* packet.
- At each intermediate node, the *S-RSV* packet attempts to reserve the selected wavelength and simultaneously configure the optical

switch. If the wavelength can be reserved, the *S-RSV* packet will be forwarded. Otherwise, as shown in Figure 6.12(b), a negative acknowledgment (*I-NAK*) packet will be sent to the destination node. This *I-NAK* packet will disconfigure the optical switch and release the wavelength already reserved by the *S-RSV* packet at each intermediate node.

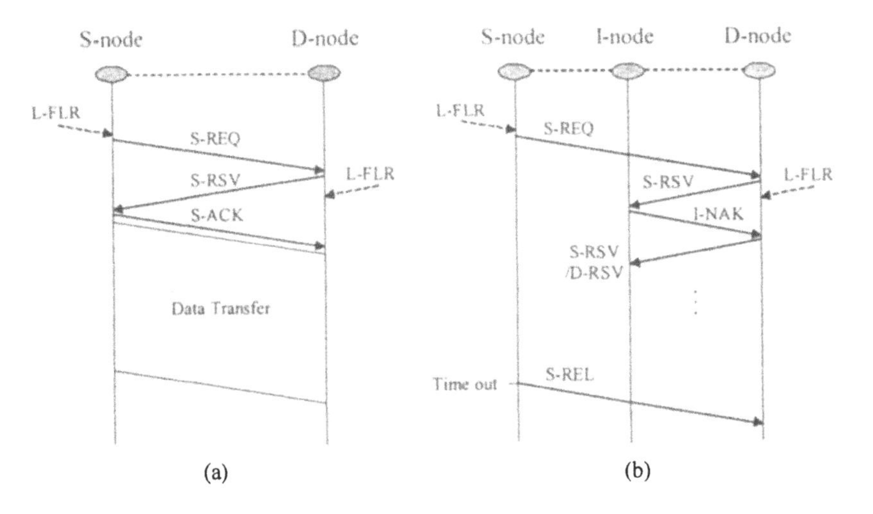

Figure 6.12 Source-initiated process: (a) successful; (b) unsuccessful.

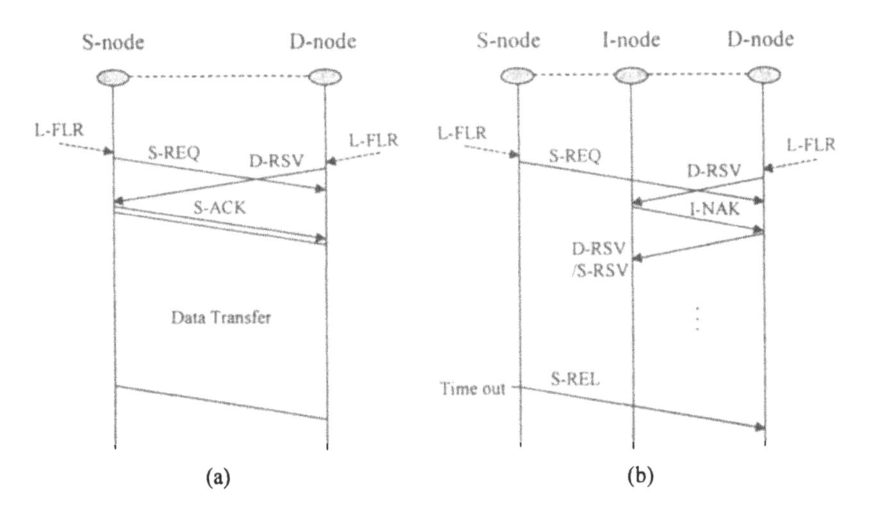

Figure 6.13 Destination-initiated process: (a) successful; (b) unsuccessful.

- When the destination node receives the *I-NAK* packet, it will attempt wavelength reservation on another available wavelength along the same route or on an available wavelength along another route, whichever route is shorter.
- If the *S-RSV* packet reaches the source node, it implies that a backup path for the disrupted connection has been established successfully, as shown in Figure 6.12(a). In response, the source node will send an acknowledgment (*S-ACK*) packet to the destination node to confirm the reservation success and then terminate its restoration process. When the destination node receives the *S-ACK* message, it will also terminate its restoration process.
- Once the destination node receives an *L-FLR* packet, it also initiates a connection restoration process. In the restoration process, if the destination node receives the *L-FLR* packet after it receives an *S-REQ* packet from the source node and sends out a *D-RSV* packet, the destination node will simply wait for the next control packet. Otherwise, it will perform a routing algorithm to decide a new route and select an available wavelength for the disrupted connection, and then send a reservation (*D-RSV*) packet to the source node along the decided route.
- At each intermediate node, the *D-RSV* packet attempts to reserve the selected wavelength and simultaneously configure the optical switch. If the wavelength can be reserved, the *D-RSV* packet will be forwarded. Otherwise, as shown in Figure 6.13(b), an *I-NAK* packet will be sent to the destination node. This *I-NAK* packet will disconfigure the local optical switch and release the wavelength already reserved by the *D-RSV* packet at each intermediate node.
- When the destination node receives the *I-NAK* packet, it will attempt wavelength reservation on another available wavelength along the same route or on an available wavelength along another route, whichever route is shorter.
- If the *D-RSV* message reaches the source node, as shown in Figure 6.13(a), it implies that a backup path for the disrupted connection has been established successfully. In response, the source node will send an *S-ACK* packet to the destination node to confirm the reservation success and then terminate its restoration process. When the destination node receives the *S-ACK* packet, it will also terminate its restoration process.

- The source node can also use a timer to control the restoration time in a way similar to that described in Section 6.6.2.

6.6.5 Multi-Initiation-Based Restoration

Consider a link failure as shown in Fig. 6.9. Suppose that it occurs on the link between node 3 and node 4. Assume that the primary path of a disrupted connection consists of six nodes with node 0 being the source node and node 5 the destination node. If source-initiated restoration is used to establish a backup path for the disrupted connection, node 3 will send a link failure (*L-FLR*) packet to the source node reversely along the primary path on the detection of the failure. However, because of the propagation delay on each link, this packet will reach the source node after a certain delay. This delay can be significantly large, especially if the failed link is closer to the destination node. For this reason, it may have a big impact on the connection restoration time or service recovery time. In dynamic restoration, each intermediate node on the primary path may have a record of information about the connection, such as the connection identifier, the destination address, and the wavelength used for the connection, during the establishment of the connection. This means that each intermediate node also has the ability to initiate a connection restoration process in the event of a link failure. Intuitively, the closer a node (at the upstream side of the failed link) is to the failed link, the earlier it can receive the *L-FLR* message and thus initiate a restoration process, which is illustrated in Figure 6.14.

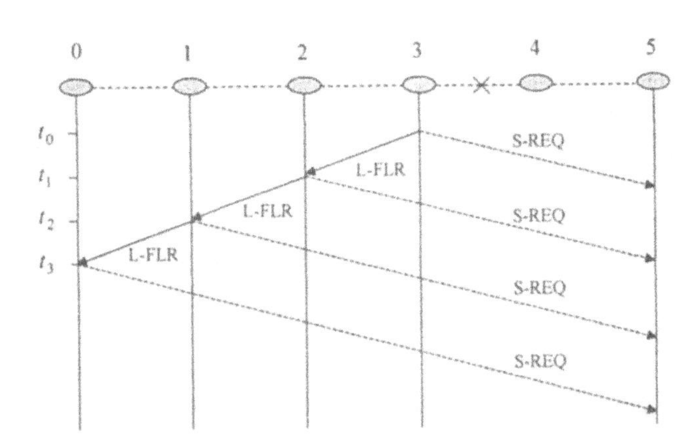

Figure 6.14 Illustration of multi-initiation-based restoration.

Suppose that node 3 detects the link failure at time t_0. Thus it can initiate a path restoration process immediately at time t_0. Subsequently, node 2, node 1, and node 0 will initiate a path restoration process after they receive the *L-FLR* message at time t_1, t_2, and t_3, respectively. Obviously, node 3, node 2, and node 1 can initiate a restoration process earlier than the source node. On the other hand, the closer a node is to the failed link, the closer it is likely to be to the destination node. For both reasons, a node closer to the failed link is likely to establish a backup path faster than the source node. Note that the *S-REQ* message in Figure 6.14 is a probe packet defined later.

Based on this observation, a multi-initiation-based restoration mechanism was proposed in [39] that allows multiple nodes, rather than only the source node or/and the destination node, on the primary path of a disrupted connection to participate in the restoration of a disrupted connection in the event of a link failure. Each of these nodes respectively initiates a restoration process on detection or notification of the link failure. In each of the restoration processes, the initiating node attempts to dynamically establish a backup path from itself to the destination node. Because multiple restoration processes can be initiated for the same disrupted connection, duplicate backup paths may be established, which is not desirable. To avoid this, the destination node serves as a coordinator among multiple restoration processes and accordingly backward reservation is used for wavelength reservation. It was shown in [39] that this restoration mechanism can significantly reduce the connection restoration time compared with the source-initiated restoration mechanism. Moreover, this restoration mechanism can be combined with different routing algorithms to achieve fast service recovery in different network scenarios.

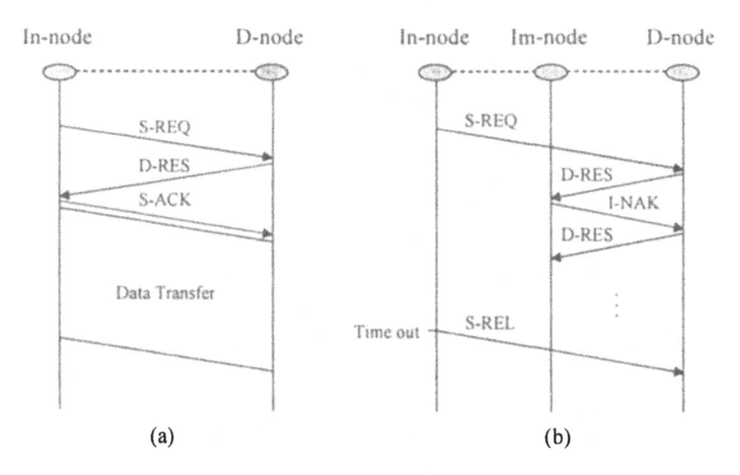

Figure 6.15 Backward reservation: (a) successful; (b) unsuccessful.

Example 6.9: A multi-initiation-based path restoration protocol based on fixed-alternate routing is briefly described as follows. For ease of exposition, we still consider the case in Figure 6.9.

- On the detection of a link failure, the end nodes of the failed link send a link failure (*L-FLR*) packet to the source and destination nodes of the disrupted connection, respectively. The *L-FLR* packet to the destination node will disconfigure the local optical switch and release the wavelength reserved for the connection at each intermediate node, whereas the *L-FLR* packet to the source node will simply inform each upstream node of the link failure.
- On the detection or notification of the link failure, the source-side end node (i.e., node 3) and each of its upstream nodes will initiate a connection restoration process, respectively.
- In each of the restoration processes, an initiating node (In-node) first performs a routing algorithm to decide on a route from itself to the destination node and then employs a backward reservation protocol to reserve an available wavelength along the decided route.

Note that if the initiating node is the source node, any available wavelength can be used for the backup path. Otherwise, the wavelength must be the same wavelength as that used for the disrupted connection unless wavelength converters are available.

- To reserve an available wavelength along the decided route, the initiating node will first send a request (*S-REQ*) packet to the destination node along the decided route, which contains information about the connection identifier, the node address, the destination address, the path list, etc. At each intermediate node (Im-node), the *S-REQ* packet collects wavelength usage information on each link of the decided route.
- When the destination node receives an *S-REQ* packet, it will select an available wavelength based on the collected information and then send a reservation (*D-RES*) packet to the initiating node along the path specified in the *S-REQ* packet. Meanwhile, it will also store the *S-REQ* packet. At each intermediate node, the *S-RES* message attempts to reserve the selected wavelength and simultaneously configure the optical switch. If the wavelength can be reserved, the *S-RES* packet is forwarded. Otherwise, a negative acknowledgment (*I-NAK*) packet will be sent to the destination node, as shown in Figure 6.15(b). The *I-NAK* packet will disconfigure the optical switch and release the wavelength already reserved by the *D-RES* packet at each intermediate node.

- When the destination node receives the *I-NAK* packet, it will reattempt wavelength reservation on another available wavelength along the same route or another route, whichever route is shorter. If there is no wavelength available within a specified time, the destination node will send a negative acknowledgment (*D-NAK*) packet to each initiating node.

Note that the destination node may receive more than one *S-REQ* packets, each arriving at a different time and containing a different path list. To reduce the connection restoration delay as far as possible, the destination node always chooses the route specified by the *S-REQ* packet that arrives first. If a subsequent *S-REQ* packet arrives, the destination node will store the packet until a successful reservation is confirmed. If the reservation is not successful, the destination node will reattempt wavelength reservation on another available wavelength along the shortest route specified in the stored *S-REQ* packets.

- If an initiating node receives a *D-RES* message, it implies that a backup path has been established from itself to the destination node successfully. In response, the node will first send an acknowledgment (*S-ACK*) packet to the destination node to confirm the restoration success. Meanwhile, if the node is the end node of the failed link, i.e., node 3, it will reconfigure the local optical switch and send an *S-ACK* packet to its upstream nodes, i.e., node 2, node 1, and node 0. If the node is not the end node of the failed link, say node 1, it will reconfigure the optical switch, and send an *S-ACK* packet to its upstream nodes, i.e., node 0, and a release (*S-REL*) packet to its downstream nodes, i.e., node 2 and node 3. The *S-ACK* packet will inform the upstream nodes of the restoration success as well as relevant information about the backup path. It will also terminate the restoration process at each upstream node. The *S-REL* packet will terminate the restoration process and meanwhile disconfigure the optical switch and release the wavelength reserved for the disrupted connection at each downstream node.

- If each initiating node receives a *D-NAK* packet, it will terminate its restoration process and meanwhile disconfigure the optical switch and release the wavelength reserved for the disrupted connection. In this case, the disrupted connection will be blocked.

- To control the restoration time, a timer can also be used at the source node in a manner similar to that described in Section 6.6.2. If the timer times out without receiving a *D-RES* packet or a *D-*

NAK packet, the source node will block the disrupted connection. Meanwhile, it will send a release (*S-REL*) packet to the destination node as well as each of the downstream nodes to terminate the restoration process. The *S-REL* packet will also disconfigure the optical switch and release the wavelength reserved for the disrupted connection at each of the initiating nodes.

6.7 Summary

Network survivability is of crucial importance in wavelength-routed WDM networks. Optical layer protection and restoration can provide fast service recovery, efficient resource utilization, protocol transparency, and some other advantages. In this chapter, we introduced a variety of basic protection and restoration schemes for the optical layer. Basically, static protection is faster in service recovery but less efficient in resource utilization whereas dynamic restoration is more efficient in resource utilization but slower in service recovery. In static protection, dedicated protection is faster and can guarantee service recovery from any failure. However, it has lower resource utilization. Shared protection is more efficient in resource utilization, but cannot handle multiple failures that occur simultaneously. On the other hand, link protection is faster in service recovery than path protection. Segment protection has several advantages over path protection. In dynamic restoration, path restoration is more efficient in service recovery than link restoration whereas link restoration has a faster restoration speed than path restoration. For static traffic, the survivable network design problem can be formulated as an integer linear programming (ILP) problem. Such ILP formulations have been known to be NP complete and therefore are only tractable for small networks. For large networks, heuristics are often used to obtain approximate solutions. In survivable routing for dynamic traffic, both a primary path and a backup path must be established for each connection request. Such survivable routing can be based on different routing strategies and protection schemes, and performed under either centralized control or distributed control. In general, static routing strategy is simpler to implement than dynamic routing strategy. Joint routing strategy can provide better network resource utilization than separate routing strategy. Distributed routing is more complicated to implement than centralized routing. In the context of dynamic restoration, the bi-initiation-based restoration mechanism is faster than both the source-initiated mechanism and destination-initiated mechanism. The multi-initiation-based restoration mechanism can reduce the restoration time compared with the source-initiated mechanism.

Problems

6.1 Why is it attractive to provide survivability at the optical layer? What are the advantages of optical-layer survivability?

6.2 What are the two basic types of survivability paradigms at the optical layer? Explain their advantages and disadvantages.

6.3 What is the difference between dedicated protection and shared protection? Describe their advantages and disadvantages.

6.4 What is the difference between path protection and link protection? Describe their advantages and disadvantages.

6.5 What is the concept of survivable routing? What is the objective of survivable routing?

6.6 What are the major requirements of survivable routing?

6.7 What is the difference between the static strategy and the dynamic strategy used in survivable routing? Give their advantages and disadvantages.

6.8 What are the separate routing strategy and the joint routing strategy? Give their advantages and disadvantages, respectively?

6.9 Is destination-initiated path restoration always faster than source-initiated path restoration? If not, give an example.

6.10 Can a multi-initiation-based restoration protocol use a routing algorithm other than fixed-alternate routing? Give an example to explain.

References

[1] James Manchester, Paul Bonenfant, and Curt Newton, "The evolution of transport network survivability," *IEEE Communications Magazine*, Aug. 1999, pp. 44–51.

[2] Ornan Gerstel and Rajiv Ramaswami, "Optical layer survivability: A services perspective," *IEEE Communications Magazine*, vol. 38, no. 3, Mar. 2000, pp. 104–113.

[3] Ornan Gerstel and Rajiv Ramaswami, "Optical layer survivability: An implementation perspective," *IEEE Journal on Selected Areas in Communications*, vol. 18, no. 10, Oct. 2000, pp. 1885–1899.

[4] Dongyun Zhou and Suresh Subramaniam, "Survivability in optical networks," *IEEE Network*, vol. 14, no. 6, Nov./Dec. 2000, pp. 16–23.

[5] S. Ramamurthy and Biswanath Mukherjee, "Survivable WDM mesh networks, part I—protection," *Proceedings of IEEE INFOCOM'99*, vol. 2, New York, pp. 744–751.

[6] S. Ramamurthy and Biswanath Mukherjee, "Survivable WDM mesh networks, part II—restoration," *Proceedings of ICC'99*, vol. 3, Vancouver, Canada, pp. 2023–2030.

[7] Sudipta Sengupta and Ramu Ramamurthy, "From network design to dynamic provisioning and restoration in optical cross-connect mesh networks: An architecture and algorithmic overview," *IEEE Network*, vol. 15, no. 4, Jul./Aug., 2001, pp. 46–54.

[8] Tsong-Ho Wu, "Emerging technologies for fiber network survivability," *IEEE Communications Magazine*, vol. 33, no. 2, Feb. 1995, pp. 58–74.

[9] Thomas E. Stern and Krishna Bala, *Multiwavelength Optical Networks: A Layered Approach*, Prentice Hall PTR, Upper Saddle River, New Jersey, May 1999.

[10] Guangzhi Li, Dongmei Wang, Charles Kalmanek, and Robert Doverspike, "Efficient distributed path selection for shared restoration connections," *Proceedings of IEEE INFOCOM'02*, vol. 1, Jun. 2002, pp. 140–149.

[11] Shengli Yuan and Jason P. Jue, "Shared protection routing algorithm for optical networks," *SPIE Optical Networks Magazine*, vol. 3, no. 3, May/Jun. 2002, pp. 32–39.

[12] Pin-Han Ho and H. T. Mouftah, "A framework for service-guaranteed shared protection in WDM mesh networks," *IEEE Communications Magazine*, vol. 40, no. 2, Feb. 2002, pp. 97–103.

[13] C. Vijaya Saradhi and C. Riva Ram Murthy, "Dynamic establishment of segmented protection paths in single and multiple-fiber WDM mesh networks," *Proceedings of SPIE Optical Networking and Communications,* vol. 4874, Aug. 2002, pp. 211–222.

[14] Dahai Xu, Yizhi Xiong, and Chunming Qiao, "Novel algorithms for shared segment protection," *IEEE Journal on Selected Areas in Communications*, vol. 21, no. 8, Oct. 2003, pp. 1320–1331.

[15] M. Alanyali and E. Ayanoglu, "Provisioning algorithms for WDM optical networks," *IEEE/ACM Transactions on Networking*, vol. 7, no.5, Oct. 1999, pp. 767–778.

[16] S. Baroni et al., "Analysis and design of resilient multifiber wavelength routed optical transport networks," *IEEE/OSA Journal of Lightwave Technology*, vol. 17, no. 5, May 1999, pp. 743–758.

[17] B. T. Doshi et al., "Optical network design and restoration," *Bell Labs Technical Journal*, vol. 4, no. 1, Jan./Mar. 1999, pp. 58–84.

[18] Hui Zang and Biswanath Mukherjee, "Path-protection routing and wavelength-assignment (RWA) in WDM mesh networks under duct-layer constraints," *IEEE/ACM Transactions on Networking,* vol. 11, no. 2, Apr. 2003, pp.248 - 258.

[19] Murari Sridharan and Arun K. Somani, "Design for upgradability in mesh-restorable optical networks," *SPIE Optical Networks Magazine*, vol. 3, no. 3, May/Jun. 2002, pp. 77–87.

[20] Schin'ichi Arakawa and Masayuki Murata, "Lightpath management of logical topology with incremental traffic changes for reliable IP over WDM networks," *SPIE Optical Networks Magazine*, vol. 3, no. 3, May/Jun. 2002, pp. 68–76.

[21] Y. Miyao and H. Saito, "Optimal design and evaluation of survivable WDM transport networks," *IEEE Journal on Selected Areas in Communications*, vol. 16, no. 7, Sep. 1999, pp. 1190–1198.

[22] D. Papadimitriou "Inference of shared risk link groups," IETF Internet draft, draft-many-inference-srlg-00.txt, work in progress.

[23] Chunsheng Xin, Yinghua Ye, Sudhir S. Dixit, and Chunming Qiao, "A joint lightpath routing approach in survivable optical networks," *SPIE Optical Networks Mag*azine, vol. 3, no. 3, May/Jun. 2002, pp. 13–20.

[24] Vishal Anand and Chunming Qiao, "Dynamic establishment of protection paths in WDM networks, part I," *Proceedings of 1999 International Conference on Computer Communications and networks (IC3N'99)*, Las Vegas, Nevada, Oct. 2000, pp. 198–204.

[25] Y. Liu, D. Tipper, and P. Siripongwutikorn, "Approximating optimal space capacity allocation by successive survivable routing," *Proceedings of IEEE INFOCOM'01*, vol. 2, Apr. 2001, pp. 699–708.

[26] G. Li, R. Doverspike and C. Kalmanek, "Fiber span failure protection in optical networks," *SPIE Optical Networks Magazine*, vol. 3, no. 3, May/Jun. 2002, pp. 21–31.

[27] Ramesh Bhandari, *Survivable Networks: Algorithms for Diverse Routing*, Kluwer Academic Publisher, Boston, 1999.

[28] D. Dunn, W. Grover, and M. MacGregor, "Comparison of k-shortest paths and maximum flow routing for network facility restoration," *IEEE Journal on Selected Areas in Communications*, vol. 12, no. 1, 1994, pp. 88–89.

[29] G. Mohan, C. Siva Ram Murthy, and Arun K. Somani, "Efficient algorithms for routing dependable connections in WDM optical networks," *IEEE/ACM Transactions on Networking*, vol. 9, no. 5, Oct. 2001, pp. 553–566.

[30] Chava Vijaya Saradhi and C. Siva Ram Murthy, "Routing differentiated reliable connections in WDM optical networks," *SPIE Optical Networks Magazine*, vol. 3, no. 3, May/Jun. 2002, pp. 50–67.

[31] K. Kompella et al., "OSPF extensions in support of generalized MPLS," IETF Internet draft, draft-ietf-ccamp-ospf-gmpls-extensions-12.txt, Oct. 2003, work in progress.

[32] K. Kompella et al., "IS-IS extensions in support of generalized MPLS," IETF Internet draft, draft-ietf-isis-gmpls-extensions-19.txt, Oct. 2003, work in progress.

[33] P. Ashwood-Smith et al., "Generalized MPLS-RSVP extensions," IETF Internet draft, work in progress, Nov. 2001.

[34] Hui Zang, Jason P. Jue, and Biswanath Mukherjee, "A review of routing and wavelength assignment approaches for wavelength-routed optical WDM

networks," *SPIE Optical Networks Magazine*, vol. 1, no. 1, Jan. 2000, pp. 47–60.

[35] J. Zheng and H. T. Mouftah, "Distributed lightpath control in wavelength-routed WDM networks," *The Handbook of Optical Communication Networks, CRC Press LIC*, Chapter 15, Boca Raton, Florida, Apr. 2003, pp. 273–286.

[36] J. Zheng and H. T. Mouftah, "A destination-initiating path restoration protocol for wavelength-routed WDM networks," *IEE Proceedings-Communications*, vol. 149, no. 1, Feb. 2002, pp. 18–22.

[37] J. Zheng and H. T. Mouftah, "A fast link restoration protocol for wavelength-routed WDM networks," *Proceedings. of 2001 IASTED International Conference on Wireless and Optical Communications (WOC'01)*, Banff, Canada, Jun. 2001, pp. 39–42.

[38] J. Zheng and H. T. Mouftah, "A bi-directional lightpath restoration mechanism for GMPLS-based WDM networks," *IEE Proceedings-Communications,* vol. 150, no.6, Dec. 2003, pp. 409–413.

[39] J. Zheng and H. T. Mouftah, "Dynamic lightpath restoration using multi-initiation mechanism for GMPLS-based WDM networks," *Broadband Networks* (to appear).

Chapter 7

IP over WDM

7.1 Introduction

The emergence of the Internet and its applications based on the Internet Protocol (IP) has opened up a new era in telecommunications. In the past decade, we have witnessed the huge success and explosive growth of the Internet, which have been driving the demand for larger bandwidth in the underlying telecommunications infrastructure. As new time-critical multimedia applications, such as Internet telephony, video conferencing, video on demand, and interactive gaming, become pervasive in the Internet, this bandwidth demand will continue to grow at a rapid rate. It has been widely considered that IP is going to be the common traffic convergence layer in telecommunications networks and that IP traffic will become the dominant traffic in the next-generation Internet. In parallel with the growth of the Internet, optical technology has also seen rapid development in the past ten years. The emergence of WDM technology has provided an unprecedented opportunity to dramatically increase the bandwidth capacity of telecommunications networks. Currently, there is no other technology on the horizon that can more effectively meet the ever-increasing demand for bandwidth in the Internet transport infrastructure than WDM technology [1]. For this reason, IP over WDM has been envisioned as the most promising network architecture for the next-generation optical Internet. To support IP over WDM, a large number of research efforts are going on to introduce more intelligence into the control plane of the underlying optical transport infrastructure and make the optical control plane more flexible, controllable, and survivable. As a result of these efforts, multiprotocol lambda switching (MPLmS) [2] and, more recently, generalized multiprotocol label switching (G-MPLS) [3–4] have been proposed by the Internet Engineering Task Force (IETF) as the control plane for dynamic provisioning and restoration of lightpaths in optical networks. Both MPLmS and G-MPLS are based on the multiprotocol label switching (MPLS) traffic engineering control plane [5–9]. In fact, MPLmS is essentially the MPLS traffic engineering control plane with extensions to address the particular characteristics of optical networks. G-MPLS further extends and generalizes MPLS to support multiple types of switching paradigms, including not only packet switching, but also TDM switching, lambda switching, and fiber switching.

This chapter introduces fundamental concepts and technologies for supporting IP over WDM networks, including various IP over WDM layered models, service models, and interconnection models that have already been proposed in the literature, and the MPLS, MPLmS, and G-MPLS control plane technologies that have already been proposed and are being standardized by the IETF. Moreover, network survivability for IP over WDM networks is discussed.

7.2 IP over WDM Layered Models

In the current Internet transport infrastructure, WDM has been mostly deployed to provide point-to-point transmission and SONET/SDH is used as the standard layer for interfacing the optical layer (or WDM layer) to the higher layers (e.g., the ATM layer and the IP layer). SONET (synchronous optical network) is a synchronous transmission system standardized by the American National Standards Institute (ANSI) and has been widely deployed in North America. SDH (synchronous digital hierarchy) is a similar system of SONET, which is standardized by the International Telecommunication Union-Telecommunication Standardization Sector (ITU-T) and has been mainly deployed in Europe and Japan. SONET/SDH has been used to provide transmission, multiplexing, and protection services in many IP transport networks. It also supports a wide range of operation, administration, and maintenance (OAM) functions as well as interoperability between equipment from different vendors. The most attractive advantages of SONET/SDH are its high-speed transmission and fast service protection capabilities. Today, commercially available SONET/SDH systems can operate at a rate up to OC-48 (2.5 Gbps) and OC-192 (10 Gbps). SONET/SDH protection mechanisms can provide a high level of service availability in the event of network failures such as fiber cuts or equipment faults, with the service restoration time less than 50 ms. However, because SONET/SDH is a synchronous transmission system and operates in circuit switching, it is more suitable for real-time applications, such as voice and video. For data applications, it may not be able to achieve efficient bandwidth utilization because of its inflexible synchronous time division multiplexing (TDM) mechanism. A more detailed description of SONET/SDH systems can be found in [10].

ATM (asynchronous transfer mode) is a networking and switching standard that was developed by the ATM forum and has been adopted by ITU-T. ATM uses fixed-size packets called cells for transferring data traffic and can support a variety of applications, such as voice, image, video, and data. An ATM cell consists of 53 bytes, with five bytes reserved for the header and 48 bytes carrying payload. This cell size provides a trade-off between the different requirements for transferring real-time traffic and data traffic. In general, a small size is preferred for real-time traffic in order to achieve a short delay whereas a large size is preferred for data traffic in order to reduce the overhead in a single packet and thus increase the efficiency in data transfer. Because of its asynchronous TDM characteristic and small size of cells, ATM has several attractive advantages over synchronous TDM systems. One of them is the flexibility in bandwidth allocation. Because the size of a cell is very small, the allocated bandwidth can range from a few kilobits per second (Kbps) to several gigabits per second (Gbps). As a result, ATM can greatly reduce bandwidth waste because of padding, resulting in improved bandwidth utilization. Another attractive advantage of ATM is the ability to provide quality-of-service (QoS) guarantees to different applications in terms of bandwidth and end-to-end delay. Even so, ATM also has some shortcomings. A major shortcoming of ATM is the 5-byte header, which introduces considerable overhead and thus greatly reduces bandwidth utilization. For a comprehensive introduction to ATM technology, the readers are referred to [11].

Because the SONET/SDH layer and the ATM layer can be used as the intermediate layers between the WDM layer and the IP layer, there are several layered models to support IP over WDM or integrate the IP layer with the WDM layer in the Internet transport infrastructure [1][12–16], as shown in Figure 7.1.

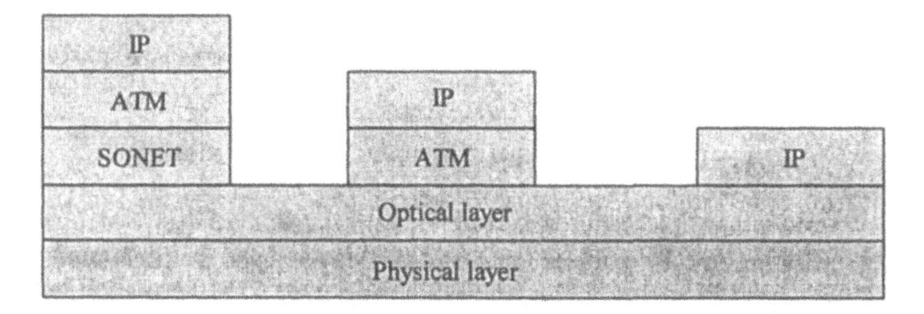

Figure 7.1 IP over WDM layered models.

7.2.1 IP over ATM over SONET/SDH over WDM

This is a commonly applied model for transporting IP traffic over WDM networks, which involves four layers: IP, ATM, SONET/SDH, and WDM. In this four-layered model, the IP layer uses the service provided by the ATM layer, while the ATM layer uses the service provided by the SONET/SDH layer, which in turn uses the service provided by the WDM layer.

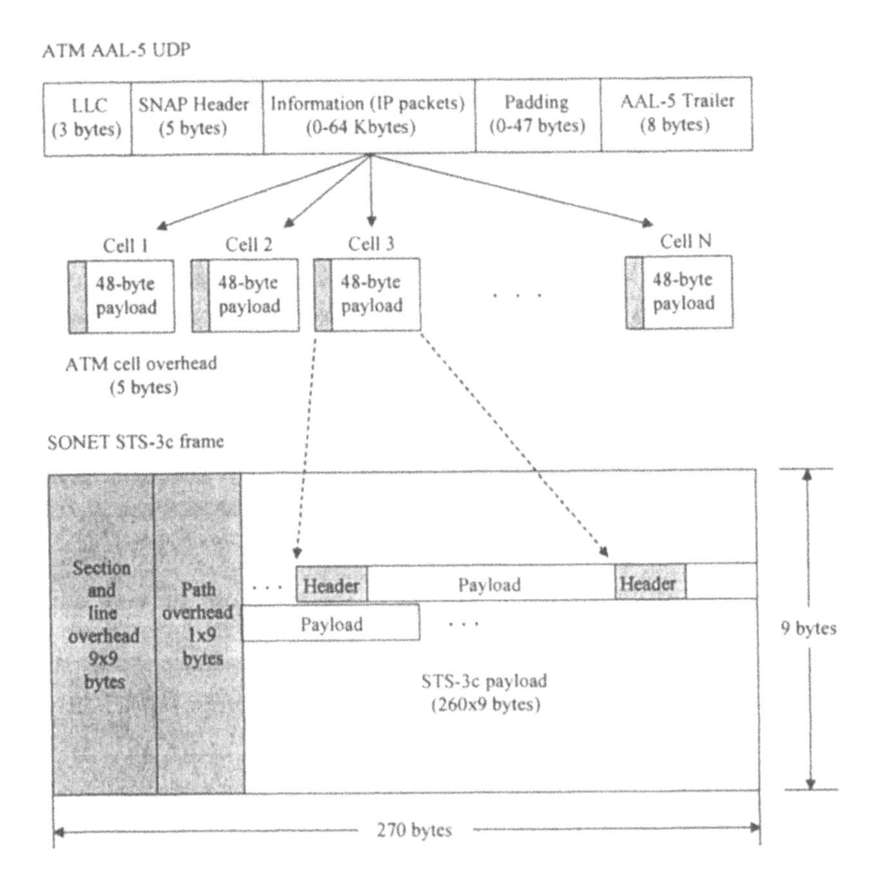

Figure 7.2 Encapsulation for IP over ATM over SONET.

The mapping for IP packets onto optical wavelengths is performed as follows. Each IP packet is first encapsulated into an ATM adaptation layer type 5 (AAL-5) protocol data unit (PDU) with multiprotocol logic link control (LLC) and subnetwork attachment point (SNAP) encapsulation. The AAL-5 PDU is then segmented into 48-byte payloads for ATM cells. The ATM cells are encapsulated into SONET/SDH frames, which are then multiplexed for transmission on WDM links. Figure 7.2 illustrates the encapsulation process for IP over ATM over SONET. This four-layered model has incorporated the significant network functions provided by all four layers, including high-speed transmission, flexible bandwidth allocation, and fast service protection. The SONET/SDH layer provides high-speed transmission at a rate up to OC-48 and OC-192 as well as high service protection capability against fiber cuts or equipment faults. The ATM layer provides flexible bandwidth allocation capability. However, this model introduces considerable bandwidth overhead, including approximately four percent SONET/SDH overhead and a much higher percentage (about 18–25% [17]) of AAL5 overhead and ATM cell overhead, which greatly decreases the data transmission efficiency. Moreover, because this model involves four layers, it greatly increases the complexity and cost in network management and operation. To increase the transmission efficiency, and reduce the network cost and complexity, it is desirable to eliminate one or more of the intermediate layers and run IP directly over WDM.

7.2.2 IP over SONET/SDH over WDM

As mentioned in Section 7.2.1, the overhead introduced by the ATM layer greatly reduces the transmission efficiency and bandwidth utilization, in particular, as the capacity of IP links grows to a rate of several gigabits per second. This has led to the motivation to eliminate the ATM layer from the four-layered model and employ a three-layered model with only the IP, SONET/SDH, and WDM layers, i.e., IP over SONET/SDH over WDM. Obviously, this model can significantly increase transmission efficiency and bandwidth utilization. Meanwhile, the high-speed transmission and fast service protection capabilities inherent in the SONET/SDH layer still remain. The shortcoming of this model is that the flexibility in bandwidth allocation with the ATM layer is also eliminated.

In this model, the mapping for IP packets into SONET/SDH frames can be performed by using the point-to-point protocol (PPP)/high-level data link control (HDLC), which has been defined in IETF RFC 2615 [18], American National Standards Institute (ASNI) SONET specifications [19–20], and ITU-T SDH Recommendation G.707 [21]. With PPP/HDLC, IP packets are first encapsulated into PPP packets, which are then framed with HDLC frames. The PPP/HDLC frames are then encapsulated into SONET/SDH frames. The format of PPP/HDLC frames is shown in Figure 7.3(a). An alternative to PPP/HDLC frames is to use the simple data link (SDL) protocol [22]. With SDL, IP packets are encapsulated into SDL frames, which are then encapsulated into SONET/SDH frames. The format of SDL frames is shown in Figure 7.3(b).

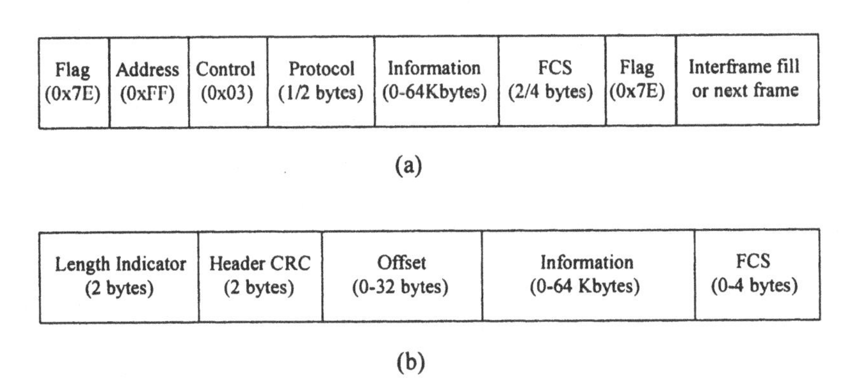

Figure 7.3 Frame formats: (a) PPP/HDLC; (b) SDL.

7.2.3 IP over WDM

A closer integration of IP and WDM is to eliminate both the SONET/SDH and ATM layers and run IP directly over WDM, resulting in a two-layered model. In this model, IP packets can be encapsulated into PPP/HDLC or SDL frames, which are then directly mapped onto optical wavelengths. Because the overhead introduced by both SONET/SDH and ATM layers has been eliminated, this two-layered model can achieve more efficient bandwidth utilization than the three-layered model. Meanwhile, it can also reduce complexity and cost in network management and operation as well. However, because of the elimination of the two intermediate layers, many of the SONET/SDH and ATM functions, such as fast protection switching and flexible bandwidth allocation, are also eliminated. For this reason, either the WDM layer or the IP layer should be enhanced to provide such functions. The emergence of the MPLS control plane technology and its extensions, MPLmS and G-MPLS, has provided better capabilities for supporting IP directly over WDM.

As a whole, the IP directly over WDM model provides a number of significant advantages over the multilayered models, such as more flexible network management, better network scalability, more efficient network operation, better traffic engineering, less complexity in management, and less cost in operation. However, this model also presents many challenges in terms of network control architecture, lightpath provisioning, network control and management, and network survivability, which are the focus of this chapter.

7.3 IP over WDM Network Model

In this section, we introduce the network model considered in many IETF documents for standardization of IP over WDM networks [15][23].

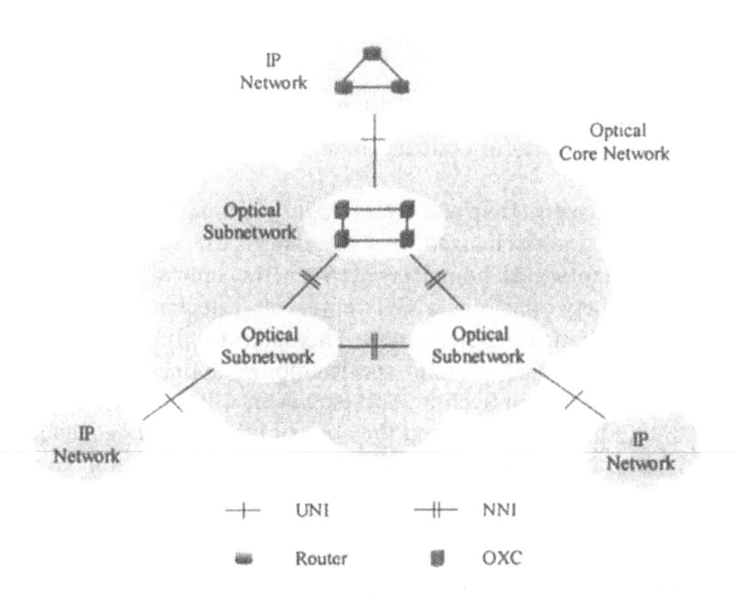

Figure 7.4 Network model for IP over WDM.

7.3.3 Network Model

The network model consists of IP client networks interconnected by dynamically established lightpaths through an optical core network, as shown in Figure 7.4. The optical core network consists of multiple subnetworks interconnected by WDM links in an arbitrary mesh topology and is incapable of processing IP packets. Each subnetwork consists of OXCs interconnected by WDM links also in an arbitrary topology. The OXCs can be either all-optical or non-all-optical. The interconnection between subnetworks is through compatible physical interfaces, with suitable O/E and E/O conversions where necessary. The IP routers that have a physical interface to the optical network are referred to as edge routers. Other client networks such as ATM networks may also be connected to the optical core network. In this network model, a lightpath must be established between a pair of IP routers before they can communicate with each other. This lightpath might have to traverse multiple subnetworks. Each subnetwork may employ a different control mechanism for dynamic provisioning and restoration of lightpaths.

7.3.2 UNI and NNI

There are two types of logical control interfaces defined in the network model shown in Figure 7.4.

- The user-network (or client-optical) interface (UNI)
- The network-network (or optical subnetwork) interface (NNI)

The difference between UNIs and NNIs mainly lies in the type and amount of control information exchanged over the interfaces. In general, the UNI represents a technological boundary between the optical network and its electronic client networks. The NNI represents a technological boundary between different optical subnetworks. The control information exchanged over UNIs and NNIs depends on the service models defined at the interfaces, which will be described in Section 7.4. However, it is theoretically possible to eliminate the difference between the control information exchanged over UNIs and over NNIs and use unified control information over both interfaces.

The logical control interfaces can be implemented with different physical control interfaces, depending on the network control architecture. For example, UNIs can be implemented with a direct interface or an indirect interface, which are described as follows.

- Direct interface: An in-band or out-of-band IP control channel may be implemented between an edge router and each OXC that it connects to. This control channel is used for exchanging signaling and routing messages between the router and the OXC(s). In this case, the edge router and the OXC(s) are viewed as peers in the control plane. The type of routing and signaling messages exchanged over the direct interface depends on the service model.
- Indirect interface: An out-of-band IP control channel may be implemented between a client network and the optical network. This control channel is used to exchange signaling and routing messages

between the control and management systems in the client network and in the optical network. In this case, there is no direct control interaction between an edge router and each OXC it connects to.

Although there are different types of control interfaces, we assume direct interfaces for both IP-optical and optical subnetwork control interactions hereafter unless otherwise stated.

7.4 IP over WDM Service Models

In this section, we introduce two service models defined over the IP-optical UNIs and the optical subnetwork NNIs for IP over WDM networks [12][15].

7.4.1 Domain Service Model

In the domain service model, the optical network primarily provides high-bandwidth connectivity to the IP client networks in the form of lightpaths. For this purpose, standardized signaling protocols are needed at the UNIs to initiate a set of optical network services, which are described as follows:

- Lightpath creation: This service allows a lightpath with specified attributes to be created between a pair of edge routers in the optical network.
- Lightpath deletion: This service allows an existing lightpath to be deleted.
- Lightpath modification: This service allows certain parameters of a lightpath to be modified.
- Lightpath status enquiry: This service allows the status of certain parameters of a lightpath to be queried by the routers that created the lightpath.

In addition to these services, a set of services for address resolution may also be available at the UNIs, which are described as follows:

- Client registration: This service allows a client to register its address and user group identifier to the optical network, which will be associated with an optical-network-administered address.
- Client deregistration: This service allows a client to deregister its address and user group identifier from the optical network.

- Address query: This service allows a client to provide another client's original address and user group identifier and get back an optical-network-administered address.
- End-system discovery: This service is used to verify local port connectivity between the optical and client devices, and it allows each device to bootstrap the control channel over the UNIs.
- Service discovery: This service allows a client to determine the static parameters of the interconnection with the optical network.

Because only a small set of services are defined at the UNIs, the signaling protocol over the UNIs only needs to exchange a few messages with certain attributes between the client IP networks and the optical network. The domain service model does not deal with the routing protocols within the optical network.

7.4.2 Unified Service Model

In the unified service model, the IP and optical networks are treated as a single network that is controlled and managed in a unified manner. As far as the control plane is concerned, an OXC in the optical network is treated just like an IP router in the IP networks. From a routing and signaling point of view, there is no difference between UNI, NNI, and any other router-to-router interface. The control plane used in this model is an MPLS-based control plane, i.e., MPLmS or G-MPLS, which will be described in Section 7.8 and Section 7.9, respectively.

In this unified model, MPLS-based signaling protocols are used to initiate a set of optical network services, which are similar to those services provided in the domain service model, that is:

- Lightpath creation: allowing a lightpath with specified attributes to be created between a pair of edge routers in the optical network
- Lightpath deletion: allowing an existing lightpath to be deleted
- Lightpath modification: allowing certain parameters of a lightpath to be modified
- Lightpath status enquiry: allowing the status of certain parameters of a lightpath to be queried by the routers that created the lightpath

In this model, however, the interconnection of the IP and optical networks is more seamless than that in the domain service model. In another word, the optical network services can be initiated in a more seamless manner compared with those in the domain service model. For example, a remote router could compute an end-to-end path across the optical network by using a routing protocol and then establish a label switched path (LSP) across the optical network by using a signaling protocol. But the edge routers must still recognize that an LSP across the optical network is a lightpath, or a conduit for multiple LSPs. Once a lightpath is established across an optical network between two edge routers, it can be advertised as a virtual link between the two routers. The bandwidth management in the unified service model is identical to that in the domain service model.

7.5 IP over WDM Interconnection Models

In this section, we introduce three interconnection models defined for IP over WDM networks [12][15][23]. The definitions of these interconnection models are mainly based on whether there is a single instance or separate instances of routing and signaling protocols for the IP and the optical networks.

7.5.1 Overlay Model

In the overlay model, there are two separate control planes or instances of routing and signaling protocols for the IP and optical networks. In other words, the routing and signaling protocols used in the IP networks are independent of those used in the optical network. The IP layer acts as a client to the optical layer, and the optical layer provides point-to-point connectivity to the IP layer. The interaction between the two layers is through the UNIs. An IP router can only see the lightpaths across the optical network while the internal topology of the optical network is invisible to the router, as shown in Figure 7.5. The lightpaths across the optical network may be either established dynamically through routing and signaling protocols or configured statically through a network management system. Moreover, the optical network can implement a registry that allows edge routers to register their IP addresses and an edge router may be allowed to query for external IP addresses in the optical network. A successful query would return the address of the egress optical port through which the external destination can be reached.

One of the advantages of the overlay model is the separation of the control planes for the IP and optical networks, which provides failure isolation, domain security, and independent evolution of technologies in both the IP and optical networks. Moreover, this hierarchical model provides better network scalability and network survivability. It has been widely considered that the overlay model is the most practical model for deployment in the short term.

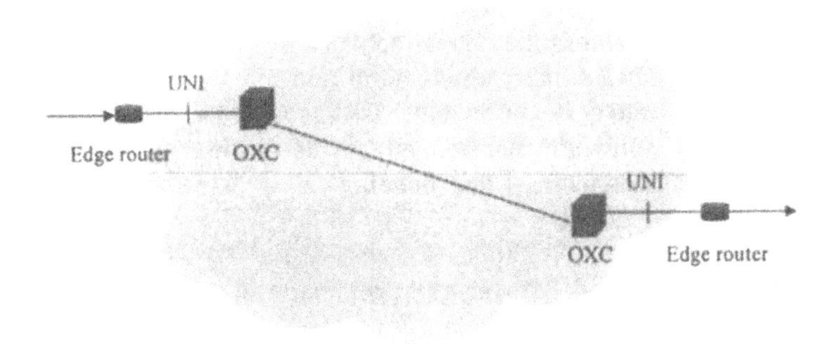

Figure 7.5 Overlay model.

7.5.2 Peer Model

In the peer model, there is only a single control plane or instance of routing and signaling protocols in both the IP and the optical networks. In other words, the routing and signaling protocols used in the IP networks are the same as those used in the optical network. A common addressing scheme is used for both the IP and the optical networks, which makes all OXCs in the optical network IP addressable and allows an IP router to see the internal topology of the optical network, as shown in Figure 7.6. Accordingly, the routers in the IP networks act as peers of the OXCs in the optical network and both routers and OXCs share a common view of the entire network topology. As far as the control plane is concerned, an OXC in the optical network is treated just as any router in the IP networks. A common interior gateway protocol (IGP), such as Open Shortest Path First (OSPF) [24] or Immediate System to Immediate System (IS-IS) [25], with appropriate extensions, may be used to exchange topology information. These extensions must be able to carry information about optical link parameters and any constraints specific to the optical network. The topology and link state information maintained at all OXCs and IP routers are identical. This allows an IP router to be able to compute an end-to-end path to another router across the optical network. Once a path is computed, an LSP can be established by using an MPLS signaling protocol, such as the resource reservation protocol with traffic engineering (RSVP-TE) [26] or the constraint-based routing label distribution protocol (CR-LDP) [27]. If the LSP is routed over the optical network, a lightpath must be established between two edge routers. This lightpath usually has a bandwidth capacity much larger than that required by this LSP. Therefore, it can accommodate not only this LSP but also other LSPs.

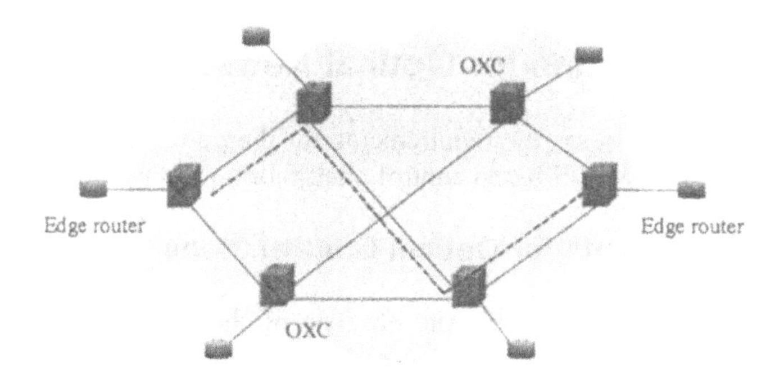

Figure 7.6 Peer model.

The peer model is more flexible and effective than the overlay model, and it allows a seamless interconnection of IP and optical networks. However, it is more complex to implement and thus is considered a long-term solution.

7.5.3 Augmented Model

In the augmented model, there are actually separate routing instances in the IP and the optical networks. Routing within the IP and optical networks is separated, each running its own routing protocol with a standard routing protocol running between them. However, both routing instances can exchange the routing information with each other. For example, external IP addresses could be assigned to OXCs and carried by the routing protocol in the optical network to allow the routing information to be shared with the IP routers in the IP networks.

In this model, the major concern is how to exchange the routing information over the UNIs. To address this problem, there are two choices that can be considered. One choice is to use an interdomain IP routing protocol, such as the border gateway protocol (BGP), which may be adapted for exchanging the routing information between the IP and optical networks. This would allow IP routers to advertise IP addresses to the optical network and to receive external IP addresses from the optical network. The other choice is to use a hierarchical routing protocol such as OSPF or IS-IS to exchange the routing information across the UNIs. The augmented model combines the advantages of the overlay model and the peer model. Compared with the peer model, the augmented mode is relatively easier for deployment in the short term.

7.6 Control Plane for Optical Networks

In this section, we discuss the requirements for the optical network control plane and introduce MPLS-based control planes for optical networks.

7.6.1 Requirements for Optical Control Plane

In general, a network control plane consists of the following four main components:

- An addressing scheme
- A communication network for exchanging routing and signaling information between network nodes
- A routing mechanism for exchanging routing information and making routing decisions
- A signaling mechanism for establishing, maintaining, and releasing optical connections

The control plane of traditional optical transport networks is implemented through network management, which has several disadvantages [2]:

- It leads to relatively slow convergence in the event of network failures (typically in minutes, or even days and months).
- It precludes the possibility of using distributed dynamic control capabilities.
- It complicates the task of internetworking equipment from different vendors, especially at the management level.
- It complicates the task of internetworking provisioning.

To support IP over WDM, the underlying optical infrastructure must be versatile, reconfigurable, and capable of providing a variety of protection and restoration mechanisms. This imposes several requirements on the optical network control plane [2], which are described as follows:

- The capability to establish lightpaths expeditiously (in seconds or even milliseconds rather than minutes, days, or months)
- The capability to support traffic engineering functions

- The capability to provide various protection and restoration mechanisms
- The capability to support interoperability between multi-vendor networks

7.6.2 MPLS-Based Optical Control Planes

To provide the capabilities required for the optical network control plane, a lot of research efforts have been already carried out and are on going to introduce more intelligence into the control plane of optical networks and make it more flexible, controllable, and survivable. There is a consensus that optical networks should make use of an IP-based control plane for dynamic provisioning and restoration of lightpaths. This is based on the practical considerations that the routing and signaling protocols developed for IP networks could be adapted for optical networks and a uniform control plane for both IP and optical networks would eliminate the complexity of controlling and managing hybrid internetworking systems. The MPLS traffic engineering control plane has successfully combined the conventional IP control plane with traffic engineering capability and has proved to be an extensible general-purpose control plane paradigm that can be used in conjunction with a variety of data planes, e.g., ATM and frame relay. This control plane can support all the capabilities that are equivalent to the requirements for the optical network control plane in the electrical domain. On the other hand, the control plane for label switch routers (LSRs) and that for OXCs have many similar requirements. Both of them involve a set of networking functions, including resource discovery, state information distribution, connection management, etc. In view of these facts, it would be very suitable to use the MPLS traffic engineering control plane as the control plane for optical networks. As a result of these efforts, MPLmS and, more recently, G-MPLS have been proposed as the optical network control plane for dynamic provisioning and restoration of lightpaths in optical networks. MPLmS is essentially the MPLS traffic engineering control plane with optical extensions. G-MPLS further extends and generalizes MPLS to support multiple types of switching paradigms, including not only packet switching, but also TDM switching, lambda switching, and fiber switching. In the next three sections, we will give a brief introduction of the MPLS, MPLmS, and G-MPLS control planes, respectively.

7.7 Multiprotocol Label Switching

Multiprotocol label switching (MPLS) has emerged as an IETF standard intended to improve the performance, scalability, and service provisioning capabilities of IP networks. A framework for MPLS was presented in [28], and the MPLS architecture was described in [29]. This networking technology is developed to address some of the limitations of conventional IP networks, in particular, the capability in traffic engineering. MPLS routing protocols provides connection-oriented services and uses LSPs to transfer IP packets. It integrates network-layer (layer 3) routing with link-layer (layer 2) switching and thus combines the flexibility of layer-3 routing with the performance and quality of service of layer-2 switching. For this purpose, MPLS decouples the routing paradigm from the forwarding paradigm, which makes the provisioning of various routing services independent of the forwarding paradigm. Meanwhile, it uses a label-swapping mechanism (e.g., ATM) for packet forwarding, which simplifies the forwarding paradigm and greatly increases the forwarding speed. Moreover, MPLS is independent of the link-layer technology. It can be based on any link-level technology, such as ATM, frame relay, and WDM. As a result, MPLS provides a number of functional capabilities over traditional IP routing and forwarding. One of the most useful applications of MPLS is traffic engineering. Other important applications of MPLS include virtual private networks (VPNs), quality of service provisioning, IP layer restoration, etc.

7.7.1 Packet Forwarding and Label Swapping

An MPLS network consists of a number of LSRs that support MPLS, as shown in Figure 7.7. The LSRs in the core of the network are called core LSRs or simply LSRs, whereas those at the edge of the network are called edge LSRs. In the MPLS network, IP packets are transferred over LSPs. An LSP typically originates at an edge LSR called the ingress-edge LSR, traverses one or more core LSRs, and terminates at another edge LSR called the egress-edge LSR. At the ingress-edge LSR, incoming packets are mapped onto LSPs with the notion of a forward equivalent class (FEC). An FEC is a set of packets that have the same attributes, such as the destination IP address prefix and the quality of service (QoS) class, and are forwarded in the same manner. Each FEC is assigned a label, which is a short fixed-length identifier used to identify the FEC, usually of local significance. All packets belonging to the same FEC are assigned the same label and are forwarded to the same LSP. At each core LSR, the labeled packets are forwarded to the next hop with a label swapping mechanism. Specifically, each LSR first uses the incoming label as an index into a label forwarding table at the LSR to get the forwarding information. This forwarding table contains the forwarding information related to each incoming label, such as the outgoing label, the outgoing port, and the next hop address. Such information is created during the establishment of each LSP. After the forwarding information is obtained, the incoming label is swapped with the outgoing label and the packets are forwarded to the next hop through the outgoing port. At the egress-edge LSR, the label is removed and the packets are forwarded with conventional IP forwarding. The packet forwarding process is also illustrated in Figure 7.7.

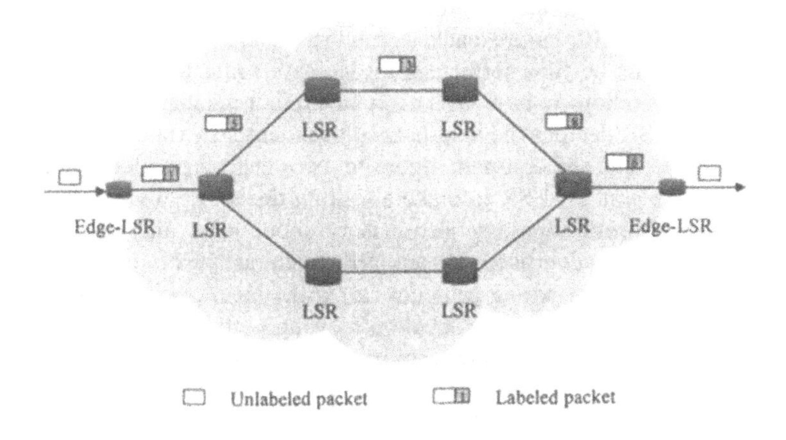

Figure 7.7 MPLS network architecture.

Label: label value (20 bits)
QoS: quality of service level (3 bits)
S: botton of stack (1 bit)
TTL: time to live (8 bits)

Figure 7.8 Format of a shim header.

Figure 7.8 gives the format of a shim header that is used to label a packet. This shim header contains 32 bits. The label field occupies 20 bits, which is used to carry the value of a label. A 3-bit field is reserved for experimental use, which has actually been used to provide different QoS levels for some applications. A single bottom of stack bit is used to indicate if a label is at the bottom of the stack and thus support the concept of label stacking. An 8-bit time to live (TTL) field is used to prevent packets from a loop in the network.

7.7.2 Routing and Path Establishment

In an MPLS network, an LSP must be established and labels must be assigned at each LSR before packets can be transferred. To establish an LSP, routing must be first performed to decide a route for the LSP, which can be either on a hop-by-hop basis or in an explicit manner. In hop-by-hop routing, each LSR decides the next hop independently by running a shortest-path algorithm. The shortest-path algorithm uses the information maintained in a routing table at the LSR to make a routing decision. The routing table contains the network topology information about links and nodes in the network, which is advertised by an IGP, such as OSPF or IS-IS. To establish the LSP, a signaling protocol called the label distribution protocol (LDP) [30] is used to assign a label at each hop. In explicit routing, the ingress-edge LSR of an LSP decides the entire route for the LSP based on the network topology and link state information maintained at the LSR. To establish the LSP, two different signaling protocols can be used to distribute label information between LSRs. One is the resource reservation protocol (RSVP) with traffic engineering extensions (RSVP-TE) [26]. The other is the LDP with traffic engineering extensions, i.e., the constraint-based LDP (CR-LDP) [27]. With explicit routing, an ingress-edge LSR can perform a constraint-based routing algorithm to compute an explicit route for each LSP, which takes into account various constraints imposed on the network in terms of physical topologies, network resources, and network policies, and can thus provide traffic engineering capability in IP networks.

7.7.3 Traffic Engineering with MPLS

With the explosive growth of the Internet, traffic engineering has become imperative in the design of the underlying Internet transport infrastructure. Traffic engineering addresses the optimization problem of network performance. It intends to achieve guaranteed quality of service, improved network resource utilization, and enhanced service recovery capability through better traffic control [7–9]. The ultimate objective is to make the network more scalable and survivable and to provide network services in a more efficient, faster, and more reliable manner.

Because of the limited capabilities of the conventional IP forwarding paradigm, traffic engineering is difficult to implement in the Internet. To address the explosive growth of Internet traffic, there have been a lot of standardization activities in the IETF and other standard organizations in recent years to develop standards for traffic engineering in IP networks. These standardization efforts have resulted in the emergence of the MPLS traffic engineering control plane, which has opened up new possibilities to address some of the limitations of the conventional IP forwarding paradigm for traffic engineering and has been able to provide the following traffic engineering capabilities:

- Topology discovery
- State information distribution
- Path selection
- Path management

Topology discovery intends to distribute relevant information about the network topology among all network nodes. This is achieved by using OSPF traffic engineering extensions (OSPF-TE) [31] or IS-IS traffic engineering extensions (IS-IS-TE) [32] to carry additional information about the network topology in their link state advertisements. The topology information is used by a constraint-based routing algorithm to compute an explicit route for LSPs.

State information distribution intends to distribute relevant information about the state of the network among all network nodes, including the link state and resource availability. This is also achieved by using OSPF-TE or IS-IS-TE to carry additional information about the state of the network in their link state advertisements. The state information is used by a constraint-based routing algorithm to compute a route for LSPs.

Path selection intends to compute an explicit route for an LSP between two edge LSRs. This is accomplished by performing a constraint-based routing algorithm, also called a constrained shortest-path-first algorithm, based on the network topology and link state information.

Path management intends to establish, maintain, and release an LSP. This is accomplished by using a signaling protocol such as CR-LDP or RESV-TE.

For more details about MPLS traffic engineering control plane, the readers are referred to [7–9][28–29][31–32].

7.8 Multiprotocol Lambda Switching

Multiprotocol lambda switching (MPLmS) is a control plane paradigm proposed by the IETF for optical networks. It is essentially the MPLS traffic engineering control plane with appropriate extensions to address the particular characteristics of optical networks. MPLmS exploits recent advances in MPLS traffic engineering control plane technology and can thus meet the requirements for the optical network control plane. As outlined in [33], MPLmS has the following advantages.

- It offers a framework for optical-layer bandwidth management and dynamic provisioning and restoration of optical channels in optical networks.
- It exploits recent advances in the control plane technology developed for MPLS traffic engineering. As a result, it obviates the need to develop a new class of control planes for optical networks and facilitates the rapid development and deployment of a new class of OXCs.
- It simplifies network administration by providing uniform semantics for network control and management in both the electrical and optical domains.

7.8.1 Analogy between MPLS and MPLmS

To use the MPLS traffic engineering control plane as the optical network control plane, an analogy should be extended between the electronic domain and the optical domain. For example, the physical fiber between a pair of OXCs represents a single link in the optical transport network topology. A wavelength or optical channel is analogous to a label, a lightpath analogous to an LSP, wavelength assignment analogous to label assignment, etc. Like MPLS, MPLmS uses the IGP extensions for traffic engineering (i.e., OSPF-TE or IS-IS-TE) with additional extensions for optical networks to distribute relevant information about the optical network topology as well as information about the wavelength availability on each fiber. An MPLS signaling protocol, such as RSVP-TE, is used to establish lightpaths. However, there are still some important differences between MPLS and MPLmS.

- There is no analogy to label merging in the optical domain, which means that an OXC cannot merge several wavelengths into one wavelength.

- There is no analogy to label stacking in the optical domain, which implies that an OXC cannot perform the equivalent operations of label pushing and popping in the optical domain because pushing and popping wavelengths cannot be implemented with current optical technologies.
- An LSR that operates in the electrical domain can potentially support an arbitrary number of LSPs with arbitrary bandwidth reservation granularities, whereas an OXC in the optical domain can only support a small number of lightpaths with coarse discrete bandwidth granularities (e.g., OC-48 and OC-192).

7.8.2 Extensions to Routing and Signaling Protocols

In an MPLmS network, a lightpath must be established between a pair of ingress and egress nodes and a wavelength must be assigned at each hop before data traffic can be transferred. Like MPLS, MPLmS supports two routing paradigms: hop-by-hop routing and explicit routing. In hop-by-hop routing, each node independently decides the next hop for a lightpath as in conventional IP networks based on the routing information maintained at the node. This information is distributed by using an existing IGP, such as OSPF or IS-IS. In explicit routing, the ingress node of a lightpath explicitly decides the entire route for the lightpath by performing a constraint-based routing algorithm. To compute an explicit route, the ingress node must have relevant information about the optical network topology and resource availability, such as the number of fibers between each pair of nodes and the number of wavelengths available on each fiber. For this purpose, MPLmS uses an existing IGP with extensions for traffic engineering, such as OSPF-TE or IS-IS-TE, to carry additional information in the link state advertisements (LSAs). For example, to support lightpath routing, optical LSAs need to be added to conventional OSPF and IS-IS. These optical LSAs include a number of new TLVs, such as link type, link media type/link resource, link ID, local interface IP address, remote IP address, traffic engineering metric, and shared risk link group TLV. By performing constraint-based routing, an ingress node can compute an explicit route for a lightpath based on various constraints imposed on the network and can thus provide traffic engineering capabilities in optical networks.

To dynamically establish, maintain, and release an explicit lightpath, a signaling protocol is needed to distribute relevant control information between adjacent nodes. For this purpose, existing RSVP-TE and CR-LDP have been proposed as the signaling protocol. Because both protocols support constraint-based routing for establishing LSPs, it is natural to extend them to support lightpath provisioning in an MPLmS network. Both RSVP-TE and CR-LDP should be extended with objects such that when used in conjunction with available network state information distributed by the routing protocols, the objects are able to provide sufficient information to establish various reconfiguration parameters for OXCs. For this purpose, extensions to RSVP-TE and CR-LDP are addressed in [34] and [35], respectively.

To support MPLmS, it is essential to have a communication mechanism to exchange control information between a pair of adjacent OXCs or between an OXC and a router. One way to implement this is to carry control traffic in-band on the same optical channels for carrying data traffic. This is simple in channel management because it does not introduce additional control channels. However, the control traffic must be demultiplexed at each OXC for processing, which will increase system complexity and affect data traffic delivery. To avoid this, an alternative way is to carry control traffic out-of-band on dedicated optical channels or via an independent IP network. For example, a default control channel can be preconfigured between each pair of adjacent OXCs or between an OXC and a router. All control traffic is exchanged on the default control channels.

7.9 Generalized Multiprotocol Label Switching

Generalized multiprotocol label switching (G-MPLS) is a multipurpose control plane technology proposed by the IETF to support multiple types of switching paradigms, including not only packet switching but also time slot switching, wavelength (or waveband) switching, and fiber switching [36]. It generalizes the MPLS traffic engineering control plane and supports not only devices that perform packet switching but also devices that perform switching in the time, wavelength, and space domains. The development of G-MPLS requires appropriate enhancements to the MPLS traffic engineering control plane to address the particular characteristics of additional types of switching paradigms. These enhancements, which are being standardized by the IETF, can be briefly summarized as follows:

- Extensions to OSPF and IS-IS routing protocols to distribute relevant information about optical resources and other attributes and constraints in optical networks (e.g., link types, bandwidth on wavelengths, and interface types)
- Extensions to RSVP-TE and CR-LDP signaling protocols to allow LSPs to be explicitly established in optical networks and other connection-oriented networks

- Introduction of the concepts of a forwarding hierarchy, link bundling, and unnumbered links to increase the routing scalability of optical networks
- Introduction of a new link management protocol to address the problems related to link management in optical networks

In this section, we give a conceptual and functional description of the G-MPLS control plane paradigm and discuss such enhancements in more detail.

7.9.1 Forwarding Hierarchy

In G-MPLS, the concept of LSRs in MPLS has been extended to include LSRs or, more precisely, LSR interfaces that forward data based on time slots, wavelengths, and physical ports or fibers. These new types of LSRs or LSR interfaces can be classified as follows:

- Packet-switch-capable (PSC) interfaces: This type of interface forwards data based on the header of the packets or cells that carry the data, such as an interface on an LSR that forwards data based on the shim header or an interface on an ATM switch that forwards data based on the ATM cell header.
- Time-division-multiplex-capable (TDM) interfaces: This type of interface forwards data based on the time slots that carry the data, such as an interface on a SONET/SDH cross-connect.
- Lambda-switch-capable (LSC) interfaces: This type of interface forwards data based on the wavelengths that carry the data, such as an interface on an OXC that operates at the level of a single wavelength or at the level of a waveband (or a group of wavelengths).
- Fiber-switch-capable (FSC) interfaces: This type of interface forwards data based on the fibers or ports that carry the data, such as an interface on an OXC that operates at the level of a single fiber or multiple fibers.

G-MPLS supports the concept of a forwarding hierarchy or LSP hierarchy, i.e., an LSP can be nested inside another LSP. In G-MPLS, the concept of LSPs has been extended to include LSPs established on different types of interfaces, such as a SONET connection and a lightpath. However, an LSP can be established only between interfaces of the same type. The LSP hierarchy exists between different types of interfaces. At the top of this hierarchy are FSC interfaces, followed by LSC interfaces, followed by TDM interfaces, followed by PSC interfaces. As a result, an LSP that originates and terminates on a PSC interface can be nested (together with other LSPs) into an LSP that originates and terminates on a TDM interface. This TDM LSP, in turn, can be nested (together with other TDM LSPs) into an LSP that originates and terminates on an LSC interface, which in turn can be nested (together with other LSPs) into an LSP that originates and terminates on an FSC interface. The LSP hierarchy in G-MPLS is illustrated in Figure 7.9.

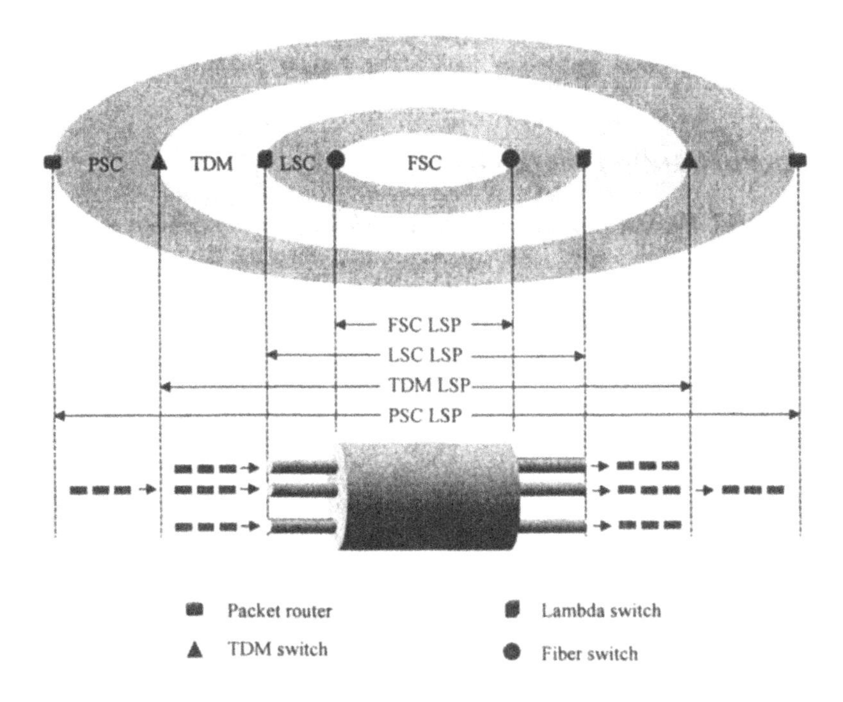

Figure 7.9 Illustration of the LSP hierarchy in G-MPLS.

To support the LSP hierarchy, the OSPF and IS-IS routing protocols should be extended to consider an LSP as a conventional link and distribute information about not only conventional links but also different types of LSPs in their link state advertisements. The LSP hierarchy addresses some of the limitations of using MPLS as the control plane for optical networks. In MPLS, the label space is relatively large, typically one million per port. Bandwidth allocation is continuous. In contrast, there are only a limited number of TDM channels and wavelengths (e.g., tens to hundreds per port) in optical networks. Bandwidth allocation is discrete. With the LSP hierarchy, the limitations in terms of the number of optical labels and the discrete optical bandwidth can be addressed. For example, a number of MPLS LSPs that enter an optical network at one node and leave the optical network at another node can be aggregated into a single optical LSP (e.g., lightpath), which can conserve the number of wavelengths used in the optical network. Moreover, each of the MPLS LSPs may have a different continuous bandwidth (e.g., 100 Mbps or 500 Mbps) and all these LSPs can share the discrete bandwidth (e.g., 2.488 Gbps) of the optical LSP. Meanwhile, the RSVP-TE and CR-LDP signaling protocols should also be extended to support the LSP hierarchy.

7.9.2 Link Bundling

In today's networks, there are only several tens or hundreds of parallel physical links (e.g., fibers, wavelengths, or TDM channels) between a pair of network nodes. To handle the explosive growth of Internet traffic, the next-generation Internet will need to deploy hundreds of parallel fibers in its underlying transport infrastructure, each carrying hundreds of wavelengths between a pair of network nodes. As a result, the total number of physical links in the network can be several orders of magnitude larger than that in an MPLS network. This requires an amount of link state information several orders magnitude larger than that in an MPLS network. To address this problem, G-MPLS introduces the concept of link bundling.

Link bundling aggregates the attributes of several parallel links of similar characteristics between a pair of nodes and assigns the aggregated attributes to a single logical link. This logical link is called a bundled link, and its physical links are called component links. A bundled link is just another kind of link defined in the OSPF or IS-IS routing protocols. The purpose of link bundling is to improve the routing scalability by reducing the amount of link state information that has to be advertised by the OSPF or IS-IS routing protocol. However, because the attributes of several links are aggregated into one single link, some of the information about the attributes of the physical links may be lost in the aggregation. For example, with a bundle of SONET links, the capacity of the link interfaces (e.g., OC-12, OC-48, or OC-192) can be advertised whereas the number of such interfaces and the time slots used may not be able to be advertised. To limit the amount of information lost in aggregation, it is necessary to restrict the type of information that can be aggregated. In G-MPLS, it is required that all component links in a bundled link be between the same pair of nodes and share some common characteristics, such as the same link type (e.g., point-to-point), the same set of resource classes (e.g., fibers, wavelengths, or TDM channels), and the same multiplexing capabilities. Although the OSPF or IS-IS routing protocol advertises a bundled link, the signaling protocol requires that individual component links be identified. To address this problem, a link management protocol has been proposed in G-MPLS, which will be introduced in Section 7.9.5.

7.9.3 Unnumbered Links

In an MPLS network, all the links are assigned IP addresses. An LSP that consists of a set of links is identified by the IP addresses of those links. The address information is used to establish the LSP by a signaling protocol. If the total number of physical links in the network increases several orders of magnitude, it becomes rather impractical to assign an IP address to each physical link because of the limited number of IP addresses. To address this problem, G-MPLS introduces the concept of unnumbered links.

Unnumbered links (or interfaces) are links (or interfaces) that do not have IP addresses assigned. For the purpose of traffic engineering, however, it is still necessary to have an alternative identifier that uniquely identifies each link in a network. In G-MPLS, this is achieved by using a tuple [router identifier, link number], where a router identifier is used to identify a particular node (or LSR) in the network and a link number is used to locally identify the outgoing link (or interface) of that particular node. Note that links are directed in the context of G-MPLS. With this type of identifiers, both the number of IP addresses used in the network and the burden in managing IP addresses can be significantly reduced, especially in an optical network with a large number of physical links. To support unnumbered links, the routing protocols must have the ability to carry information about unnumbered links. This ability has already been included in OSPF-TE and IS-IS-TE. In addition, the signaling protocols must also have the ability to indicate unnumbered links in the signaling. This requires extensions to the RSVP-TE and CR-LDP signaling protocols, because neither protocol provides support for unnumbered links.

7.9.4 Extensions to Routing and Signaling Protocols

To support multiple types of switching paradigms, G-MPLS introduces the concepts of a forwarding hierarchy, link bundling, and unnumbered links. As mentioned in the previous sections, to support these concepts the routing protocols must be able to distribute more information about different types of LSPs, bundled links, and unnumbered links than that which traditional OSPF and IS-IS routing protocols can provide, such as link types and bandwidth. Such information will be used by a constraint-based routing algorithm to compute an explicit route for an LSP. For this purpose, G-MPLS requires appropriate extensions to OSPF-TE and IS-IS-TE. For details of the extensions, the readers are referred to [37] and [38].

To support multiple types of switching paradigms, G-MPLS extends some basic functions of the RSVP-TE and CR-LDP signaling protocols and also introduces some additional functionality, which can be outlined as follows:

- A new label request format to include non-PSC characteristics
- Labels for non-PSC interfaces
- Support for waveband switching
- Label suggestion by the upstream for optimization purposes
- Label restriction by the upstream to support optical constraints
- Rapid failure notification
- Bidirectional LSP establishment with contention resolution
- Explicit routing with explicit label control

G-MPLS signaling protocols support bidirectional LSP establishment, which is a requirement of many optical network service providers. A bidirectional LSP has the same traffic engineering requirements in both directions. In MPLS, LSPs are unidirectional. To establish a bidirectional LSP with RSVP-TE or CR-LDP, two unidirectional LSPs in opposite directions must be established independently. This has a number of disadvantages, such as longer setup delay, more control overhead, and larger blocking probability [36]. With bidirectional LSPs, both are established with a single set of Path/Request and Resv/Mapping signaling messages. This reduces the setup delay and limits the control overhead to the same number of messages as a unidirectional LSP.

G-MPLS signaling protocols also support rapid failure notification, which is one of the major requirements for network survivability. In the event of a network failure, a node that detects the failure should be able to notify the nodes that are responsible for the restoration of the disrupted connections rapidly without unnecessary delay at intermediate nodes. In G-MPLS, a NOTIFY message has been added to RSVP-TE for this purpose, which is used to notify nonadjacent nodes of a network failure. However, this has not been defined in CR-LDP extensions. This NOTIFY message does not replace the ERROR message in RSVP in that it can notify any node other than the immediate upstream or downstream nodes of a network failure. For more details about RSVE-TE extensions and CR-LDP extensions, the readers are referred to [34] and [35].

7.9.5 Link Management Protocol

Link management is a collection of useful functionality for link provisioning and fault localization between two adjacent nodes. In G-MPLS, a link management protocol (LMP) [39] has been designed to provide four basic functions:

- Control channel management
- Link property correlation
- Link connectivity verification
- Fault localization

Control channel management

The control channel management function is used to establish and maintain connectivity between two adjacent nodes. In G-MPLS networks, at least one bidirectional control channel is needed for communication between two adjacent nodes, which is used to exchange control information including routing, signaling, and management information. The control channel does not have to use the same physical medium as the data links. Instead, it can use a separate wavelength or fiber, an Ethernet link, and even a separate IP network. The decoupling of the control channel from the data links results in a lack of correlation between them and thus requires control channel management in terms of link provisioning and fault localization. The control channel management is implemented by using a Hello protocol that consists of two phases: a negotiation phase and a keep-alive phase. The negotiation phase allows negotiation of some basic Hello protocol parameters, like the Hello frequency. The keep-alive phase consists of a fast, lightweight Hello message exchange.

Link property correlation

The link property correlation function is used to provide the correlation of link properties (e.g., link identifiers, protection mechanisms, and priorities) between two adjacent nodes. This is implemented by using the LinkSummary message in LMP. A link property exchange mechanism allows for dynamically changing link properties. It allows for adding data links to a bundled link, changing a link protection mechanism, changing port identifiers, or changing component link identifiers in a bundled link.

Link connectivity verification

The link connectivity verification function is used to verify the physical connectivity of data links (e.g., component links of a bundled link) as well as to exchange the link identifiers that will be further used in the RSVP-TE and CR-LDP signaling protocols.

Fault localization

The fault localization function is used to rapidly locate link failures. It can also be used to support some local protection and restoration mechanisms. Logically, fault localization occurs only after a fault is detected.

Note that the former two functions are mandatory whereas the latter two are optional. For more details of LMP, the readers are referred to [39].

7.10 Survivability in IP over WDM Networks

Survivability of IP over WDM networks is a critical problem for the next-generation optical Internet. From a layered perspective, network survivability can be provided at different network layers, not only at the optical layer but also at the higher layers, such as the SONET/SDH layer and the IP layer. A lot of research has been carried out for provisioning survivability at the WDM layer and the IP layer, respectively. To provide better survivability for IP over WDM networks, multilayer survivability has also been proposed and received much attention in recent years [40–44]. In this section, we first introduce various survivability paradigms for WDM and IP networks, and then discuss multilayer survivability for IP over WDM networks.

7.10.1 Survivability in WDM Networks

As we already discussed in Chapter 6, there are two basic survivability paradigms for the optical layer: static protection and dynamic restoration.

Static protection

In static protection, spare network resources are reserved during the establishment of each connection. In the event of a network failure, the disrupted network services are recovered by using the reserved network resources. Static protection can be further classified into path protection and link protection or into dedicated protection and shared protection, which are described as follows.

- Path protection: A backup path is reserved for the primary path on an end-to-end basis during the setup of each connection.

- Link protection: A backup path is reserved around each link of the primary path during the setup of each connection.
- Dedicated path protection: A backup path is reserved for the primary path during the setup of each connection. The wavelength reserved on each link of the backup path is dedicated to that backup path (e.g., 1+1 or 1:1 protection).
- Shared path protection: A backup path is reserved for the primary path during the setup of each connection. However, the wavelength reserved on each link of the backup path can be shared with other backup paths (e.g., 1:N or M:N protection).
- Dedicated link protection: For each link of the primary path, a backup path and a wavelength are reserved around that link during the setup of each connection. The wavelength reserved on each link of the backup path is dedicated to that backup path.
- Shared link protection: For each link of the primary path, a backup path and a wavelength are reserved around that link during the setup of each connection. However, the wavelength reserved on each link of the backup path may be shared with other backup paths.

Dynamic restoration

In dynamic restoration, no spare network resources are reserved. The network must search dynamically for spare network resources available in the network to recover the disrupted network services after a network failure occurs. Dynamic restoration can be further classified into path restoration and link restoration.

- Path restoration: The source and destination nodes of each connection that traverses the failed link search dynamically for a backup path on an end-to-end basis.
- Link restoration: The end nodes of the failed link search dynamically for a backup path around the link for each connection that traverses the failed link.

Because of their different characteristics, static protection and dynamic restoration have different advantages and disadvantages. In general, static protection is fast in service recovery but inefficient in resource utilization whereas dynamic restoration is efficient in resource utilization but slow in service recovery. The restoration time is usually hundreds of milliseconds (e.g., 200 ms), whereas the protection switching time is usually less than 100 ms (e.g. 50 ms). Link protection and restoration provide faster service recovery, whereas path protection and restoration provide higher resource utilization. Dedicated protection is very fast in service recovery and can guarantee service recovery from any failure. However, it is inefficient in resource utilization. In contrast, shared protection is more efficient in resource utilization. However, it cannot handle multiple failures that share the same reserved network resources and occur simultaneously.

7.10.2 Survivability in IP Networks

In traditional IP networks, each router dynamically maintains a routing table by running a routing protocol such as OSPF or IS-IS to distribute the routing information throughout the network. For each destination node, the routing table may have several entries, which are associated with a set of least-cost paths to the destination node and are computed based on the network topology and resource information maintained at the router. The path cost can be defined as a function of link cost, which is measured in terms of hop count, link delay, etc. Each of the entries specifies the outgoing port to the next hop of a corresponding path. Whenever there is a change in network topology, a topology update message is advertised to each router in the network, which will compute a set of new paths for each destination node based on the updated topology and resource information and then update the related entries in the routing table. With the routing table, each packet is routed from the source node to the destination node on a hop-by-hop basis. At each intermediate node, the router inspects the destination IP address, looks up the routing table to determine the outgoing port for the next hop, and then forwards the packet to the next hop. Because of this, network survivability in IP networks is achieved by rerouting through the convergence of the routing information after the detection of a failure and is essentially best-effort in nature.

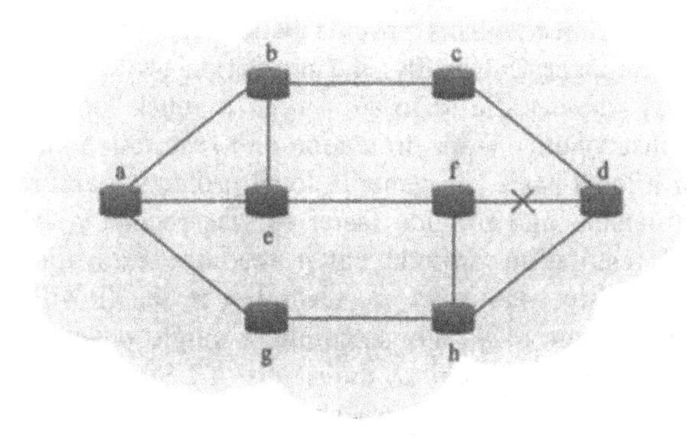

Figure 7.10 Rerouting in IP networks

Figure 7.10 illustrates how the IP layer handles a link failure by rerouting. Suppose that router *a* is sending packets to router *d* using an optimal path *a-e-f-d* and there is a link failure between router *f* and router *d*. Once router *f* detects the failure, it will first decide on an alternate path to router *d*, e.g., path *f-h-d*, based on the routing table and then forward the packets to router *h*, where the packets will further be forwarded to router *d*. At the same time, router *f* will send a topology update message to each router in the network to notify that the link between router *f* and router *d* has failed. Once router *a* receives this message, it will compute a set of new paths to router *d* and update its routing table. Then it will send packets using a new optimal path, e.g., *a-b-c-d*, to router *d*.

Survivability at the IP layer has several advantages, such as the ability to find the optimal routes through the network and the ability to provide a finer granularity of protection [42]. However, service recovery at the IP layer is usually slower, on the order of tens of seconds to a few minutes. The main reason is that the network topology usually takes up to several minutes to converge in the event of a link failure, including the time for detecting the failure and the time for propagating the topology update message. As a result, a router may make a decision based on incomplete or inaccurate topology and resource information, which would lead to a nonoptimal routing decision and thus result in slower service recovery.

The introduction of MPLS has provided the possibility of fast service recovery in IP networks by using fast rerouting based on the concepts of explicit routing and LSPs. Similar to the WDM layer, survivability mechanisms at the IP/MPLS layer can be classified into two basic categories: protection and restoration. In general, protection mechanisms provide faster service recovery but are inefficient in resource utilization whereas restoration mechanisms provide better resource utilization but are slow in service recovery. Usually, the protection switching time is under 100 ms (50 ms) whereas the restoration time is much longer (hundreds of seconds or milliseconds). Both protection and restoration can be provided on a global or a local basis. In general, local protection and restoration are simpler to implement and provide faster service recovery, whereas global protection and restoration provide better resource utilization and higher restoration probability. However, survivability at the IP/MPLS layer also has some shortcomings [42]. For example, a single physical failure (e.g., fiber cut or node fault) may lead to thousands of LSPs being disrupted. To handle this kind of network failures at the IP/MPLS layer, thousands of recovery processes are needed, resulting in a large amount of MPLS signaling overhead in the network. For this reason, service recovery at the IP/MPLS layer is not scalable.

7.10.3 Multilayer Survivability in IP over WDM Networks

In IP over WDM networks, protection and restoration capabilities are available at both the IP/MPLS layer and the WDM layer. In the event of a network failure, service recovery can be provided at either one of the layers or both layers. Because of the different technologies used at the IP/MPLS and WDM layers, however, both layers have their own characteristics in service recovery. Basically, the IP/MPLS layer provides a finer granularity of protection (e.g., at a packet level or a LSP level) than the WDM layer. However, service recovery at the IP/MPLS layer is slower and less efficient than the WDM layer. The recovery time at the IP layer is usually on the order of tens of seconds to several minutes. Moreover, the IP/MPLS layer cannot detect physical failures such as a fiber cut or a node fault in the network. In contrast, the WDM layer provides faster and more efficient service recovery than the IP/MPLS layer. The recovery time at the WDM layer is on the order of tens to hundreds of milliseconds. However, it can only provide a coarser granularity of protection (e.g., at a wavelength level or fiber level) and cannot detect a failure at the IP/MPLS layer. Usually, a failure at the WDM layer can be recovered at either the WDM layer or the IP/MPLS layer but a failure at the IP/MPLS layer can only be recovered at the IP/MPLS layer. Because of this, it is desirable to combine the survivability capabilities of both layers to provide better network survivability for different network services. For this purpose, it is very important to coordinate the service recovery at the IP/MPLS and WDM layers in an effective manner. Otherwise, the network would suffer from inefficient resource utilization and even from a failure in service recovery.

For a physical failure at the WDM layer, there are generally two basic strategies for service recovery at the IP/MPLS and WDM layers: parallel and sequential. With the parallel strategy, a service recovery process is activated at both IP/MPLS and WDM layers on the detection of the failure. However, because the service recovery at the WDM layer is faster than that at the IP/MPLS layer, it is usually unnecessary to activate the service recovery process at the IP/MPLS layer if the failure can be handled at the WDM layer. On the other hand, parallel service recovery at both layers is difficult to coordinate, which might lead to inefficient resource utilization and even result in a failure in service recovery. With sequential strategy, a service recovery process is activated at the WDM layer and the IP/MPLS layer in a sequential manner. The control complexity is less than that with the parallel strategy. The major problem is how to escalate the failure detected at the WDM layer to the IP/MPLS layer.

An effective way to implement the coordination between the WDM and IP/MPLS layers is to use a hold-off timer as proposed in [44]. A hold-off timer is used to control the interval between the instant when the service recovery process at one layer is activated to the instant when the service recovery process at the other layer is activated after the detection of the failure. The purpose is to allow a certain amount of time for the service recovery at one layer to take effect and avoid parallel service recovery at both IP/MPLS and WDM layers. For this purpose, the initial value of the hold-off timer should be large enough to allow the service recovery at one layer to complete, either successfully or unsuccessfully. If the service recovery at one layer is successful, there is no need to activate the service recovery process at the other layer. Otherwise, the hold-off timer will time out, and then the other layer will be notified of the recovery result and the service recovery process at that layer will be activated.

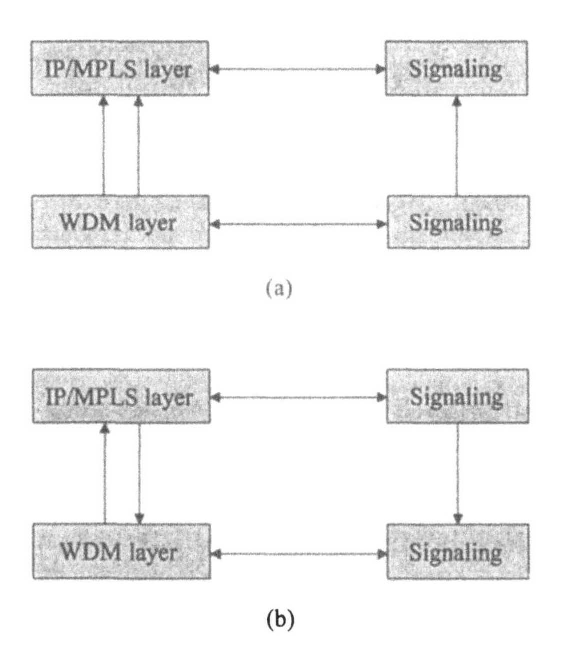

Figure 7.11 Escalation strategies: (a) bottom-up; (b) top-down.

There are two basic escalation strategies for coordinating multi-layer service recovery: bottom-up and top-down. With the bottom-up strategy, the service recovery starts at the WDM layer and the failure is escalated to the IP/MPLS layer on the timeout of the hold-off timer, as shown in Figure 7.11(a). When the WDM layer detects a failure, it will activate its own service recovery process to recover the disrupted services. If the services cannot be recovered within a specified maximum recovery time, the WDM layer will notify the IP/MPLS layer of the recovery result and the service recovery process at the IP/MPLS layer will be activated. With the top-down strategy, the service recovery starts at the IP/MPLS layer and the failure is escalated to the WDM layer on the timeout of the hold-off timer, as shown in Figure 7.11(b). When the WDM layer detects a failure, it will first notify the IP/MPLS layer of the failure and the IP/MPLS layer will activate its own service recovery process. If the IP/MPLS layer cannot recover the disrupted services, it will escalate the failure to the WDM layer. In general, the top-down strategy provides higher capacity efficiency than that of the bottom-up strategy. However, because one failure in the WDM layer may lead to many failures in the IP/MPLS layers, the top-down strategy is more complex to implement and is slower in service recovery than the bottom-up strategy.

7.11 Summary

The integration of IP and WDM has been envisioned as the most promising solution for the next-generation Internet. In this chapter, we introduced various layered models, service models, and interconnection models that have already been proposed in the literature for supporting IP over WDM in the Internet transport infrastructure. Among all different IP over WDM layered models, IP directly over WDM has been considered the most efficient and cost-effective model in terms of bandwidth utilization and network cost and complexity. For the interconnection of IP and WDM layers, the overlay model uses separate instances of routing and signaling protocols for the IP and optical networks and is the most practical model for deployment in the short term. The peer model allows a seamless interconnection of IP and optical networks but is not practical in the short term. The augmented model combines the advantages of the overlay model and the peer model, and is more practical for deployment in the short term than the peer model. On the other hand, the support for IP over WDM imposes some requirements on the optical network control plane. To meet these requirements, MPLmS and G-MPLS have been proposed by the IETF as the optical network control planes for dynamic provisioning and restoration of lightpaths in optical networks. Both MPLmS and G-MPLS are based on the MPLS traffic engineering control plane with appropriate extensions to MPLS routing protocols (i.e., OSPF-TE and IS-IS-TE) and signaling protocols (i.e., RSVP-TE and CR-LDP) to address the particular characteristics of optical networks. As a result, both MPLmS and G-MPLS can meet the requirements for the optical network control plane. In particular, G-MPLS generalizes MPLS and supports the concepts of a forwarding hierarchy, link bundling, and unnumbered links, which largely improves the routing flexibility and scalability in the integration of IP and WDM and can provide better traffic engineering capability.

In addition, survivability of IP over WDM networks was discussed. In IP over WDM networks, service recovery can be provided at both the IP/MPLS layer and the WDM layer. Basically, the IP/MPLS layer provides a finer granularity of protection than the WDM layer but the service recovery is slower and less efficient. The WDM layer provides faster and more efficient service recovery than the IP/MPLS layer but the protection granularity is coarser. Because of this, it is desirable to combine the survivability capabilities of both layers in order to provide better network survivability for different network services. For this purpose, it is very important to coordinate the service recovery at the IP/MPLS and WDM layers in an effective manner. In this context, the parallel and sequential recovery strategies were discussed. The bottom-up and top-down escalation strategies for coordinating multilayer service recovery were also discussed.

Problems

7.1 What are the most attractive advantages of SONET/SDH technology and ATM technology?

7.2 What is the main difference between the user-network interface (UNI) and the network-network interface (NNI) in the IP over WDM network model?

7.3 What are the main differences between different IP over WDM interconnection models: overlay model, peer model, and augmented model? Explain their advantages and disadvantages.

7.4 What are the major requirements for optical network control planes? Why can MPLS-based control planes be used for optical networks?

7.5 What is the most significant characteristic of MPLS? What are the most useful applications of MPLS?

7.6 Describe the packet forwarding and label swapping mechanisms in MPLS.

7.7 What are the routing protocols and signaling protocols used in MPLS? Explain their main functions.

7.8 Give the analogies and differences between MPLS and MPLmS.

7.9 What are the major switching paradigms that G-MPLS can support?

7.10 Why are extensions to MPLS routing protocols and signaling protocols needed to support G-MPLS for optical networks?

7.11 What are the advantages and disadvantages of optical-layer survivability and IP-layer survivability?

7.12 Why is it important to effectively coordinate the service recovery at the IP and WDM layers?

References

[1] Nasir Ghani, Sudhir Dixit, and Ti-Shiang Wang, "On IP-over-WDM integration," *IEEE Communications Magazine*, vol. 38, no. 3, Mar. 2000, pp. 72–84.

[2] Daniel Awduche and Yakov Rekhter, "Multiprotocol lambda switching: Combining MPLS traffic engineering control with optical crossconnects," *IEEE Communications Magazine*, vol. 39, no. 3, Mar. 2001, pp. 111–116.

[3] Ayan Banerjee et al., "Generalized multiprotocol label switching: An overview of routing and management enhancements," *IEEE Communications Magazine*, vol. 39, no. 1, Jan. 2001, pp. 144–150.

[4] Ayan Banerjee et al., "Generalized multiprotocol label switching: An overview of signaling enhancements and recovery techniques," *IEEE Communications Magazine*, vol. 39, no. 7, Jul. 2001, pp. 144–151.

[5] Grenville Armitage, "MPLS: the magic behind the myths," *IEEE Communications Magazine*, vol. 38, no. 1, Jan. 2000, pp. 124-131.

[6] Arun Viswanathan et al., "Evolution of multiprotocol label switching," *IEEE Communications Magazine*, vol. 36, no. 5, May 1998, pp. 165–173.

[7] George Swallow, "MPLS advantages for traffic engineering," *IEEE Communications Magazine*, vol. 37, no. 12, Dec. 1999, pp. 54–57.

[8] Daniel O. Awduche, "MPLS and traffic engineering in IP networks," *IEEE Communications Magazine*, vol. 37, no. 12, Dec. 1999, pp. 42–47.

[9] Xipeng Xiao et al., "Traffic engineering with MPLS in the Internet," *IEEE Network*, vol. 15, no. 2, Mar./Apr. 2001, pp. 28–33.

[10] Walter J. Goralski, *SONET/SDH*, 3rd Edition, McGraw-Hill, New York, 2002.

[11] Rainer Handel, Manfred N. Huber, and Stefan Schroder, *ATM Networks: Concepts, Protocols, and Applications*, 3rd Edition, Addison-Wesley, Harlow, England, 1998.

[12] Antonio R. Moral, Paul Bonenfant, and Murali Krishnaswamy, "The optical Internet: Architectures and protocols for the global infrastructure of tomorrow," *IEEE Communications Magazine*, vol. 39, no. 7, Jul. 2001, pp. 152–159.

[13] Paul Bonenfant and Antonio Rodriguez-Moral, "Framing techniques for IP over fiber," *IEEE Network*, vol. 15, no. 4, Jul./Aug. 2001, pp. 12–18.

[14] Marco Listanti, Vincenzo Eramo, and Roberto Sabella, "Architectural and technological issues for future optical Internet networks," *IEEE Communications Magazine*, vol. 38, no. 9, Sep. 2000, pp. 82–92.

[15] Bala Rajagopalan et al., "IP over optical networks: architectural aspects," *IEEE Communications Magazine*, vol. 38, no. 9, Sep. 2000, pp. 94–102.

[16] Chadi Assi et al., "Optical networking and real-time provisioning: An integrated vision for the next-generation Internet," *IEEE Network*, vol. 15, no. 4, Jul./Aug. 2001, pp. 36–45.

[17] J. Anderson et al., "Protocols and architectures for IP optical networking," Bell Labs Technical Journal, vol. 4, no. 1, Jan./Mar. 1999, pp.105–124.

[18] A. Malis and W. Simpson, "PPP over SONET/SDH," IETF RFC 2615, Jun. 1999.

[19] ANSI T1X1.5/2001-062, "Synchronous optical network (SONET)–basic description including multiplex structure, rates and formats," Rev. Draft, Jan. 2001.

[20] ANSI T1X1.5/2000-192R1, "Synchronous optical network (SONET)–payload mappings," Rev. Draft, Oct. 2000.

[21] ITU-T Rec. G. 707, "Network node interface for the synchronous digital hierarchy (SDH)," Oct. 2000.

[22] B. T. Doshi et al., "A simple data link protocol for high speed packet networks," *Bell Labs Technical Journal*, vol. 4, no. 1, Jan./Mar. 1999, pp. 85–104.

[23] Bala Rajagopalan et al., "IP over optical networks: a framework," IETF RFC 3717, Mar. 2004.

[24] J. Moy, "OSPF version 2," IETF RFC 2328.

[25] D. Oran, "OSI IS-IS intra-domain routing protocol," IETF RFC 1142.

[26] D. Awduche et al., "RSVP-TE: Extensions to RSVP for LSP Tunnels," IETF RFC 3209, Dec. 2001.

[27] B. Jamoussi et al., "Constraint-Based LSP Setup using LDP," IETF RFC 3212, Jan. 2002.

[28] R. Callon et al., "A framework for multiprotocol label switching," IETF Internet draft, draft-ietf-mpls-framework-05.txt, Sep. 1999, work in progress.

[29] E. Rosen, A. Viswanathan, and R. Callon, "Multiprotocol label switching architecture," IETF RFC 3031, Jan. 2001.

[30] L. Andersson *et al.*, "LDP specification," IETF RFC 3036, Jan. 2001.

[31] K. Ishiguro and T. Takada, "Traffic engineering extensions to OSPF Version 3," IETF Internet draft, draft-ietf-ospf-ospfv3-traffic-01.txt, Aug. 2003, work in progress.

[32] Henk Smit and Tony Li, "IS-IS extensions for traffic engineering," IETF Internet draft, draft-ietf-isis-traffic-05.txt, Aug. 2003, work in progress.

[33] D. O. Awduche et al., "Multi-Protocol Lambda Switching: Combining MPLS traffic engineering control with optical cross-connects," IETF Internet draft, draft-awduche-mpls-te-optical-02.txt, Jul. 2002.

[34] L. Berger, "GMPLS Signaling: Resource ReserVation Protocol-Traffic Engineering (RSVP-TE) Extensions" IETF RFC 3473, Jan. 2003.

[35] P. Ashwood-Smith, L. Berger, "GMPLS Signaling: Constraint-Based Routed Label Distribution Protocol (CR-LDP) Extensions," IETF RFC 3472, Jan. 2003.

[36] Eric Mannie et al., "Generalized Multi-Protocol Label Switching Architecture," IETF Internet draft, draft-ietf-ccamp-gmpls-architecture-07.txt, May 2003, work in progress.

[37] K. Kompella et al., "IS-IS extensions in support of generalized MPLS," IETF Internet draft, draft-ietf-isis-gmpls-extensions-19.txt, Oct. 2003, work in progress.

[38] K. Kompella et al., "OSPF extensions in support of generalized MPLS," IETF Internet draft, draft-ietf-ccamp-ospf-gmpls-extensions-12.txt, Oct. 2003, work in progress.

[39] J. Lang et al., "Link management protocol (LMP)," IETF Internet draft, draft-ietf-ccamp-lmp-10.txt, Aug. 2003, work in progress.

[40] Yinghua Ye, Sudhir Dixit, and Mohamed Ali, "On joint protection/restoration in IP centric DWDM-based optical transport networks," *IEEE Communications Magazine*, vol. 38, no. 6, Jun. 2000, pp.174–183.

[41] Yinghua Ye et al., "A simple dynamic integrated provisioning/protection scheme in IP over WDM networks," IEEE Communications Magazine, vol. 39, no. 11, Nov. 2001, pp. 174–182.

[42] H. Zhang and A. Durresi, "Differentiated multi-layer survivability in IP over WDM networks," *Proc. of 8th IEEE/IFIP Network Operations and Management Symposium (NOMS'02)*, Florence, Italy, Apr. 2002.

[43] Peter Laborczi and Tibor Cinkler, "IP over WDM configuration with shared protection," *SPIE Optical Networks Magazine*, vol. 3, no. 5, Sep./Oct. 2002, pp. 21–33

[44] Robert Doverspike and Jennifer Yates, "Challenges for MPLS in optical network restoration," *IEEE Communications Magazine*, vol. 39, no.2, Feb. 2001, pp. 89–96.

Chapter 8

Future Trends in Optical Networks

8.1 Introduction

The explosive growth of the Internet combined with emerging high-speed applications, such as Internet telephony, video conferencing, and video on demand, is imposing a huge demand for network bandwidth on the underlying telecommunications infrastructure. Traditional broadband telecommunications networks, such as ATM networks, are no longer capable of meeting such a huge bandwidth demand. Fiber-optics technology has brought about a revolution in telecommunications networks. Optical fibers have proved to be an excellent physical transmission medium with a number of advantages over traditional transmission media, such as huge transmission bandwidth (nearly 50 Tbps), low signal attenuation (about 0.2 dB/km), low error bit rate (typically 10^{-12}), low signal distortion, low power requirement, low space requirement, and low cost [1]. A single single-mode fiber can potentially provide nearly 50 Tbps bandwidth, which is about four orders of magnitude larger than the electronic transmission rates of a few gigabits per second (Gbps). The emergence of wavelength division multiplexing (WDM) technology has provided an unprecedented opportunity to exploit the huge bandwidth inherent in optical fibers. WDM allows multiple optical channels over a single fiber, each operating at a rate of a few gigabits per second, which can greatly increase the usable bandwidth of an optical fiber and thus meet the ever-increasing bandwidth demand of network users. With recent advances in enabling technologies, WDM systems capable of supporting up to 160 OC-192 (10 Gbps) channels are commercially available and products with more optical channels are expected to come into the commercial market soon. Therefore, WDM technology has been considered the most promising technology for meeting the huge bandwidth demand in the telecommunications infrastructure and optical networks employing WDM technology have become the most promising network infrastructure for next-generation telecommunications networks [2].

WDM technology is currently being deployed by several network providers mainly for point-to-point transmission. Because of the limit of electronic processing speed, this may result in an electronic processing bottleneck at a network node and thus largely affect network performance. To overcome this bottleneck, it is desirable to incorporate optical switching and routing functions into each network node to realize an all-optical WDM network. On the other hand, many emerging high-speed applications also require the telecommunications infrastructure to provide better network performance and quality of service by employing optical switching and routing technologies and all-optical communication. For these reasons, a significant amount of research from academia, industry, and standardization organizations has been motivated in the last few years to move WDM from a point-to-point transmission technology toward a networking technology.

Optical networking will not only improve network performance but also have a great impact on the way in which the network is designed. Although the commercial availability of advanced optical devices such as EDFAs, OADMs, and OXCs has paved the way for optical networking, a number of network design issues still need to be solved in order for optical networking to become a reality. These design issues include virtual topology design, routing and wavelength assignment, network control and management, and network survivability, just to name a few. Over the last few years, we have seen considerable research and development progress on optical WDM networks. A variety of effective solutions have been proposed to address these design issues. A number of experimental prototypes and testbeds have already been and are currently being developed and built by network and service providers. These have been evidenced by a large number of papers and reports published in the literature. Despite all this effort and progress, however, all-optical networks have not yet been truly realized. Optical networks with wavelength routing and switching capabilities have been deployed only in limited situations, mainly in the backbone networks [4]. The vision of all-optical WDM networks as well as the integration of optical networking and the Internet will continue to present new challenges and opportunities and to drive research, development, and commercialization for optical networking [2–6]. In this chapter, we will briefly discuss some of the future trends in optical networking as well as the major challenges that need to be addressed in order to realize the promise of all-optical WDM networks.

8.2 IP over WDM

The integration of IP and WDM will continue to be a focus in optical networking. As mentioned in chapter 7, the Internet is growing at a very fast speed and IP traffic will become the dominant traffic in future telecommunications networks. To deal with the continued growth of Internet traffic and support various emerging high-speed Internet applications, such as Internet telephony, video conferencing, and video on demand, the underlying telecommunications infrastructure must be capable of providing not only sufficient bandwidth but also quality of service and fault tolerance. IP over WDM has been envisioned as the most promising solution for meeting such challenges. Although there are several layered network models for supporting IP over WDM, such as IP over ATM over SONET/SDH over WDM, IP over SONET/SDH over WDM, and IP directly over WDM, IP directly over WDM is considered the most efficient model because it can provide a number of significant advantages over the other models, such as more flexible control, better scalability, more efficient operation, better traffic engineering, less complexity in management, and less cost in operation. However, this model also presents a number of critical issues, in terms of interface models, service models, network control, and fault management. Although these issues have been studied widely over the last two years and many solutions have been proposed in the literature, the practical deployment of IP directly over WDM networks is still not mature and needs further effort. Meanwhile, a practical concern in the transition to IP over WDM is that SONET/SDH and ATM networks have been deployed widely in today's telecommunications networks with a large investment in SONET/SDH and ATM equipment. SONET/SDH systems have several attractive advantages such as high-speed transmission and fast service protection capabilities. ATM networks also have several attractive advantages such as flexible bandwidth allocation and support for quality of service. In view of these facts, the transition to the IP directly over WDM model should be carried out in a smooth and cost-effective way. A practical strategy is to employ a multilayered model as the short-term solution, with the IP directly over WDM model being the long-term solution.

On the other hand, WDM technology is evolving from a circuit-switching technology toward burst-switching and packet-switching technologies. Both burst-switching and packet-switching technologies have a number of significant advantages over optical circuit switching, such as finer bandwidth granularity and efficient bandwidth utilization, which can better support the integration of IP and WDM. In the next two sections, we will give a brief introduction to optical packet switching and optical burst switching, respectively.

8.3 Optical Packet Switching

In this book, we have focused on wavelength-routed WDM networks, which are characterized by circuit switching or wavelength routing. In such networks, a lightpath must be first established for a connection request before data traffic can be transferred. The wavelength resources reserved for the lightpath cannot be used for other connection requests until the lightpath is released. For this reason, wavelength-routed WDM networks are more suitable for voice traffic. For data traffic, the time for transferring a packet may possibly be much shorter than the time to establish a lightpath, which may largely reduce bandwidth utilization. With the continued growth and popularity of Internet applications, it is expected that data traffic, which is bursty in nature, will surpass voice traffic and dominate future telecommunications networks [2]. To deal with the bursty characteristic of data traffic, it is desirable to employ a switching technology that is not only fast in data transmission but also efficient in bandwidth utilization. Optical packet switching (OPS) is a promising switching technology that is capable of meeting such challenges [7].

Conceptually, OPS is similar to its electronic counterpart used for traditional IP networks. In electronic packet switching, the basic switching unit is an IP packet. An IP packet consists of two parts: header and payload. The header contains information for routing, such as the IP addresses of the source node and the destination node, whereas the payload carries data traffic. At each IP router, an IP packet is forwarded to the next hop using the routing information contained in the packet header. This information is used as an index to the routing table maintained at the router to determine the next hop. An optical packet in OPS also consists of a header and a payload. The header may be processed either electronically or optically, whereas the packet payload is switched entirely in the optical domain. As a result, optical packet switching has a number of advantages over optical circuit switching, such as faster switching speed, finer switching granularity, and more efficient bandwidth utilization [7]. However, this technology is still at the laboratory stage. A number of critical issues still need to be solved before optical packet-switched networks can be deployed commercially, such as switch architecture, contention resolution, and packet synchronization.

One critical issue in OPS is the packet contention problem, which can greatly affect network performance in terms of packet loss ratio, average packet delay, and network throughput. Packet contention occurs at the time when two or more packets enter a packet switch and attempt to leave the switch from the same output port at the same time. In electronic packet switches, contention can be resolved by using the store-and-forward mechanism, in which packets in contention are temporarily stored in an electronic buffer, such as random-access memory (RAM), and are sent out once the output port becomes available. In optical packet switches, however, there are no optical buffers similar to electronic RAMs. Instead, fiber delay lines (FDL), i.e., fixed-length fibers, are usually used as buffers to store packets in the optical domain. However, because a packet entering a delay line will have to leave the line from the other end of the line after a fixed and limited amount of time, optical delay lines are not flexible in resolving packet contentions.

In general, there are three methods for contention resolution in OPS:

- Optical buffering
- Deflection routing
- Wavelength conversion

In optical buffering, optical delay lines are used to buffer packets in the optical domain. In deflection routing, packets in contention are routed and forwarded to some other links, actually using the links as the buffers. In wavelength conversion, packets in contention are forwarded to the same output port on a different wavelength. Each of these methods has its own advantages and disadvantages. For example, wavelength conversion is the most efficient method because it does not introduce additional delay to those packets in contention. However, wavelength converters are expensive, which would largely increase the cost of implementation. Deflection routing is the least costly method because it does not need any additional hardware and only needs to forward the packets in contention to some other links. However, it would introduce additional delay to those packets in contention and would reduce network throughput as well. Optical buffering has a medium cost in implementation because it needs to use optical delay lines as the buffers. However, it may also introduce additional delay to those packets in contention, depending on the packet length. In practice, we can use any combination of these three methods to achieve better contention resolution in optical packet switching.

Another critical issue in OPS is the packet synchronization problem. To perform OPS or forward a packet to the next hop, an optical switch must be able to recognize the header and the end of each packet. This requires bit-level synchronization and fast clock recovery. On the other hand, in optical packet-switched networks, packets enter a node or an optical packet switch on different input ports at different times. All the packets on different input ports must be aligned before they are switched through the switch fabric. In some situations, packet alignment may also be required on the output ports before packets leave a node to compensate the time jitter that occurs inside the switch. To achieve such alignment, synchronization resolution needs to be provided at both the input ports and the output ports, resulting in the input and output synchronization stages in an optical packet switch.

Optical packet-switched networks can be generally classified into the following two categories:

- Slotted (or synchronous)
- Unslotted (or asynchronous)

In optical slotted networks [8–9], time is divided into fixed-length slots, which are synchronized on a global basis. All packets are of the same length and each packet is transmitted in a single time slot. All packets in the same slot are destined for the same destination. Because all packets are aligned after they enter and before they leave a switch, packet contention in the switch fabric can be minimized. However, this would result in additional cost and more complex switch architecture because of the required synchronization stages. In contrast, unslotted networks allow packets to have variable lengths. The switch architecture is simpler, but packet contention would be much higher than that in slotted networks. For an overview of OPS and a survey of different switch fabrics, the readers are referred to [7–12].

8.4 Optical Burst Switching

As discussed in the previous section, OPS has a number of attractive advantages, such as faster speed, finer granularity, and more efficient bandwidth utilization, that have made it become a promising switching technology for the next-generation optical Internet. However, this technology is still a dream at the current stage. Optical packet switch prototypes are still in the laboratories. Commercial deployment of OPS in the short term is still limited by several obstacles in enabling technologies in terms of switch architectures, packet synchronization, contention resolution, etc.

An alternative to OPS is optical burst switching (OBS) [13-14]. In OBS, the basic switching unit is a burst of data or simply a burst, which is of a variable length that can range from one packet to several packets. A burst is usually assembled at an ingress node by aggregating several packets that may come from a single or multiple users in the same or different access networks and are destined for the same egress node. To deliver a burst of data, a control packet is sent first to establish a connection or to reserve the bandwidth or wavelength and configure the switches for the connection. It contains relevant information for routing, quality of service (QoS), and other purposes. This control packet is then followed by the data burst without getting an acknowledgment for the connection establishment. At each intermediate node, the control packet is processed electronically while the data burst is switched optically. The control packet will reserve the bandwidth or wavelength on an outgoing link for the data burst for the duration from the instant the burst enters the node to the instant the burst leaves completely. If the control packet fails to reserve the bandwidth, the data burst can be delayed optically for a certain time by using fiber delay lines. If there is no wavelength available even after the delay, the data burst will be dropped. If the control packet can reserve the bandwidth, it will be forwarded to the next hop. As soon as the burst passes through a link, the bandwidth or wavelength reserved for the burst on that link will be released either automatically or by an explicit release packet. This allows different bursts to share the bandwidth of the same wavelength on a link in a time-multiplexing manner, and thus increases the bandwidth utilization. OBS can support reliable burst transmission. To achieve this, a negative acknowledgment packet can be sent to the ingress node to inform of the failure in the case of a reservation failure. Then the ingress node can retransmit the control packet and the data burst at a later time.

Optical burst switching has several major characteristics that are different from optical circuit switching and OPS. OBS uses one-way reservation protocols such as tell-and-go (TAG) [15–16], whereas optical circuit switching uses two-way reservation protocols such as tell-and-wait (TAW) [17–18]. OBS uses data bursts of variable lengths as the basic switching unit, which can range from one packet to several packets and thus result in a smaller control overhead per data unit. On the other hand, the control packet and the data burst are less tightly coupled in both space and time than in OPS. In space, OBS uses out-of-band signaling, which means that the control packet and data burst are sent over separate optical or wavelength channels. In time, the control packet is sent first, followed by the data burst after an offset time at the ingress node. By choosing the offset time at the ingress node to be larger than the total processing time of the control packet along the route, one can eliminate the need for a data burst to be buffered at each intermediate node to wait for the control packet to be processed. Alternatively, an OBS protocol may choose not to use any offset time at the ingress node. Instead, the data burst must go through a fixed delay that is no shorter than the maximum time needed to process the control packet at an intermediate node.

OBS combines the advantages of wavelength routing and OPS. It provides a trade-off between coarse-granularity wavelength routing and fine-granularity optical OPS. In OBS, there is no need for buffering and electronic processing of data burst as in optical circuit switching. In addition, OBS can achieve efficient bandwidth utilization by reserving required bandwidth only for the duration in which data transmission is actually needed. From the perspective of implementation, OBS is more feasible than OPS to be deployed in the near term. For these reasons, OBS has received a lot of attention in recent years and has been extensively studied in the literature. At the present stage, however, it is still a premature technology. Several critical issues need to be solved before it can be deployed commercially. The major challenges in burst switching include the design of optical burst switches, burst switching protocols, and contention resolution without optical buffering. For more details and recent studies on optical burst switching, the readers are referred to [10][13–14][19].

8.5 Optical Metro Networks

As we have seen, the backbone networks in telecommunications infrastructure have experienced a significant evolution in the past several years with the deployment of a large amount of optical fiber links as well as various WDM systems. This has greatly changed and continues to change the picture of the backbone networks: The bandwidth capacity has been dramatically increased, and various networking functions are being provisioned. Compared with the backbone networks, however, the metro and access networks have not changed significantly. In particular, the access networks, which serve residential and small business users, have become a bandwidth bottleneck in today's telecommunications infrastructure. For example, the local subscriber lines for telephone and cable television are still using twisted pairs and coaxial cables, respectively. Most residential connections to the Internet still use dial-up modems operating at a low speed on twisted pairs. With the growth of Internet traffic, in particular the emergence of high-speed applications, the demand for bandwidth in the metro and access networks is also increasing significantly. To meet such bandwidth demand and support emerging applications, the metro and access networks must also be updated to provide not only sufficient bandwidth capacity but also a rich variety of network services. However, the requirements for the backbone, metro, and access networks are different because of their different characteristics. For the backbone networks, which have a high level of traffic aggregation, we are more concerned with bandwidth capacity, and traffic protection and restoration. For the metro and access networks, which have a reduced level of traffic aggregation, we are more concerned with the flexible provisioning of a richer variety of network services. This has presented new challenges for network designers and service providers.

Metro networks, also called metropolitan area networks, cover large metropolitan areas that span geographical distances of several tens to a hundred kilometers and interconnect the access networks to the backbone networks. Today, most metro networks employ SONET/SDH-based rings, which are based on single-mode fibers (i.e., using a single wavelength channel) and time division multiplexing (TDM) technology. With TDM, low-rate traffic streams (e.g., OC-3 and OC-12) are aggregated into a high-rate traffic stream (e.g., OC-48 and OC-192) through SONET add/drop multiplexers (ADMs). Although TDM is well suited for voice traffic, it is not scalable and efficient in bandwidth utilization and is thus unsuitable for highly dynamic data traffic. Consequently, traditional SONET/SDH rings have proven insufficient to deal with the surging data traffic growth and various emerging applications. There is an urgent need for new solutions that can provide bandwidth scalability and achieve high service flexibility in a cost-effective manner. With the emergence of WDM technology, it is natural to evolve the traditional SONET/SDH rings to WDM SONET/SDH rings. Such rings logically provide multiple traditional SONET/SDH rings, each operating on a particular wavelength. To achieve high bandwidth utilization in such rings, traffic grooming [20–27] must be performed to aggregate low-rate traffic to high-rate traffic in a cost-effective manner. This requires a new type of optical devices, i.e., optical add/drop multiplexers (OADMs), that are able to selectively bypass some wavelengths and at the same time drop the other wavelengths from a single optical signal or fiber. Although WDM SONET/SDH rings can be used as an effective solution to providing high bandwidth in the metro networks in the near term, this is not a long-term solution because the TDM-based infrastructure is not flexible and scalable. As data traffic continues to grow at a rapid speed, its bursty and unpredictable nature will become more and more dominant in future metro network traffic. To deal with such "burstiness," it is desirable to have a metro network architecture that is scalable and flexible and is capable of exploiting the full advantages of WDM technology. For this purpose, the next-generation metro networks may use WDM mesh topologies [28–29].

8.6 Optical Access Networks

Access networks, including local area networks, cover the "last mile" of the telecommunications infrastructure, typically spanning geographical distances of about 1 to 10 km, and connect service providers to individual homes and businesses. As the demand for bandwidth in access networks increases with emerging high-speed applications, the bandwidth bottleneck in the telecommunications infrastructure has shifted to the "last mile." For this reason, it becomes imperative to provide more bandwidth in access networks to support a variety of network services, such as telephone, Internet access, distance learning, high-definition television (HDTV), video on demand (VoD), and video conferencing. Without doubt, optical technology is the technology of choice for solving the bandwidth and service provisioning problems in access networks. The challenges mainly lie in the physical-layer or component-enabling technologies. To make a variety of services economically viable, access networks must be implemented at low cost. Today, several different solutions are being deployed in access networks to support emerging high-speed services, such as digital subscriber loops (DSLs) and hybrid fiber coax (HFC) [30]. With future advances in enabling technologies, the next-generation access networks are expected to bring the fibers to the building (FTTB) or to the home (FTTH), enabling high-speed applications accessible in user premises at a cost comparable to today's DSL and HFC technologies.

Optical access networks can be implemented with different technologies. The simplest way is to use a point-to-point topology, in which point-to-point fiber links connect a central office to individual user premises. This solution is simple to implement but is expensive because of the large amount of fibers to be deployed. An alternative way is to use an active star topology, in which a curb switch is installed close to a group of individual users to aggregate or distribute traffic between the individual users and the central office. This topology is more cost effective in terms of the fibers deployed. However, the curb switch is an active device that requires electronic power and maintenance. To reduce the maintenance cost, passive optical networks (PONs) are very attractive and have received a lot of attention for optical access networks [30]. In a passive optical network, a single fiber is used to connect a central office to a passive optical device such as an optical star coupler or an optical splitter, which distributes the optical signal to the curb (i.e., FTTC) or to the homes (i.e., FTTH). Passive optical network technology has several attractive advantages over other technologies. For example, passive optical networks provide higher bandwidth and allow for a longer distance (over 20 km) between a central office and user premises compared to that (about 5.5 km) with DSL. Passive optical networks use passive optical devices, which provide higher reliability, relieve maintenance, and do not need to be powered. Moreover, the fiber infrastructure itself is transparent to bit rates and protocol formats, which allows for easier update without changing the infrastructure itself. Optical access networks can be based on either WDM technology or TDM technology. At the present stage, however, WDM technology is still cost-prohibitive for an access network. TDM technology is a more cost-effective and practical technology [30]. In the future, we may envision WDM technology being deployed widely in access networks. In such deployment, two practical factors should be considered. First, the deployment must be compatible with traditional access technologies, such as Ethernet. Second, the deployment must be able to provide significant benefits over existing access solutions in order to be commercially viable. In fact, Ethernet passive optical networks (EPON), which combine the low-cost point-to-multipoint optical infrastructure with low-cost high-bandwidth Ethernet, are emerging as an attractive solution to the "last mile" problem for the next-generation access network.

Problems

8.1 What are the most attractive advantages of SONET/SDH technology and ATM technology?

8.2 Why is optical circuit switching unsuitable for data traffic?

8.3 What are the most critical issues in optical packet switching?

8.4 Describe three solutions to contention resolution in optical packet switching. What are their advantages and disadvantages?

8.5 Explain the major characteristics of optical burst switching.

8.6 What is the offset time in optical burst switching? How can the offset time be implemented at an intermediate node?

8.7 Explain the main differences between slotted networks and unslotted networks in terms of contention resolution and packet synchronization.

8.8 Why are passive optical networks attractive for access networks? Explain their advantages over other technologies.

References

[1] Biswanath Mukherjee, *Optical Communication Networks*, McGraw-Hill, New York, 1997.

[2] Biswanath Mukherjee, "WDM optical communications networks: Progress and challenges," *IEEE Journal on Selected Areas in Communications*, vol. 18, no. 10, Oct. 2000, pp. 1810–1824.

[3] Aysegul Gencata, Narendra Singhal, and Biswanath Mukherjee, "Overview of optical communication networks: current and future trends," *The Handbook of Optical Communication Networks*, CRC Press LLC, Chapter 1, Boca Raton, Florida, 2003, pp. 1–28.

[4] M. Yasin and Akhtar Raja, "Evolution of optical networks architecture," *The Handbook of Optical Communication Networks*, CRC Press LLC, Chapter 2, Boca Raton, Florida, 2003, pp. 28–44.

[5] Biswanath Mukherjee and Hui Zang, "Survey of State-of-the-art," *Optical WDM Networks: Principles and Practice*, Kluwer Academic Publishers, Chapter 1, Boston, 2000, pp. 3–24.

[6] Imrich Chlamtac and Jason P. Jue, "Optical WDM networks: future vision," *Optical WDM Networks: Principles and Practice*, Kluwer Academic Publishers, Chapter 16, Boston, 2000, pp. 343–350.

[7] Shun Yao, B. Mukherjee, and Sudhir Dixit, "Advances in photonic packet switching: An overview," *IEEE Communications Magazine*, vol. 38, no. 2, Feb. 2000, pp. 84–94.

[8] Viktoria Elek, Andrea Fumagalli, and Gosse Wedzinga, "Photonic slot routing: A cost-effective approach to designing all-all-optical access and metro networks," *IEEE Communications Magazine*, vol. 39, no. 12, Dec. 2001, pp. 164–172.

[9] Imrich Chlamtac et al., "Scalable WDM access network architecture based on photonic slot routing," *IEEE/ACM Transactions on Networking*, vol. 7, no. 1, Feb. 1999, pp. 1–9.

[10] L. Xu, H. G. Perros, and G. Rouskas, "Techniques for optical packet switching and optical burst switching," *IEEE Communications Magazine*, vol. 39, no. 1, Jan. 2001, pp. 136–142.

[11] S. Yao et al., "All-optical packet switching for metropolitan area networks: opportunities and challenges," *IEEE Communications Magazine*, vol. 39, no. 3, Mar. 2001, pp. 142–148.

[12] Chunming Qiao and Myungsik Yoo, "A taxonomy of switching techniques," *Optical WDM Networks: Principles and Practice, Kluwer Academic Publishers*, Chapter 5, Boston 2002, pp. 103–125.

[13] Chunming Qiao and Myungsik Yoo, "Optical burst switching (OBS)—a new paradigm for an optical Internet," *Journal of High Speed Networks*, vol. 8, no. 1, 1999, pp. 69–84.

[14] S. Verma, H. Chaskar, and R. Ravikanth, "Optical burst switching: A viable solution for terabit IP backbone," *IEEE Network*, vol. 14, no. 6, Nov. 2000, pp. 48–53.

[15] E. Varvarigos and V. Sharma, "The ready-to-go virtual circuit protocol: a loss-free protocol for multigigabit networks using FIFO buffers," *IEEE/ACM Transactions on Networking*, vol. 5, no. 5, Oct. 1997, pp. 705–718.

[16] I. Widjaja, "Performance analysis of burst admission-control protocols," *IEE Proceedings-Communications*, vol. 142, no. 1, Feb. 1995, pp. 7–14.

[17] P. E. Boyer and D. P. Tranchier, "A reservation principle with applications to the ATM traffic control," *Computer Networks and ISDN Systems*, vol. 24, 1992, pp. 321–334.

[18] Jonathan S. Turner, "Managing bandwidth in ATM networks with bursty traffic," *IEEE Network*, vol. 6, no. 5, Sep. 1992, pp. 50–58.

[19] Myungsik Yoo, Chunming Qiao, and Sudhir Dixit, "Optical burst switching for service differentiation in the next-generation optical Internet," *IEEE Communications Magazine*, vol. 39, no. 2, Feb. 2001, pp. 98–104.

[20] E. H. Modiano, "Traffic grooming in WDM networks," *IEEE Communications Magazine*, vol. 39, no. 7, Jul. 2001, pp. 124–129.

[21] A. A. M. Saleh and J. M. Simmons, "Architectural principles of optical regional and metropolitan access networks," *IEEE Journal of Lightwave Technology*, vol. 17, no. 12, Dec. 1999, pp. 2431–2448.

[22] A. L. Chiu and E. H. Modiano, "Traffic grooming algorithms for reducing electronic multiplexing costs in WDM ring networks," *IEEE Journal of Lightwave Technology*, vol. 18, no. 1, Jan. 2000, pp. 2–12.

[23] X. Zhang and C. Qiao, "An effective and comprehensive approach for traffic grooming and wavelength assignment in SONET/WDM rings," *IEEE/ACM Transactions on Networking*, vol. 8, no. 5, Oct. 2000, pp. 608–617.

[24] J. Wang et al., "Improved approaches for cost-effective traffic grooming in WDM ring networks: ILP formulations and single-hop and multihop connections," *IEEE Journal of Lightwave Technology*, vol. 19, no. 11, Nov. 2001, pp. 1645–1653.

[25] R. Dutta and G. N. Rouskas, "On optimal traffic grooming in WDM rings," *IEEE Journal on Selected Areas in Communications*, vol. 20, no. 1, Jan. 2002, pp. 110–121.

[26] O. Gerstel, R. Ramaswami, and G. H. Sasaki, "Cost-effective traffic grooming in WDM rings," *IEEE/ACM Transactions on Networking*, vol. 8, no. 5, Oct. 2000, pp. 618–630.

[27] Keyao Zhu and Biswanath Mukherjee, "Traffic grooming in an optical WDM mesh network," *IEEE Journal on Selected Areas in Communications*, vol. 20, no. 1, Jan. 2002, pp. 122–133.

[28] Muriel Medard and Steven Lumetta "Architectural issues for robust optical access," *IEEE Communications Magazine*, vol. 39, no. 7, Jul. 2001, pp. 116–122.

[29] Nasir Ghani, "Metropolitan networks: trends, technologies, and evolutions," *SPIE Optical Networks Magazine*, vol. 3, no. 4, Jul./Aug. 2002, pp. 7–13.

[30] G. Kramer and G. Pesavento, "Ethernet passive optical networks (EPON): building a next-generation optical access network," *IEEE Communications Magazine*, vol. 40, no. 2, Feb. 2002, pp. 66–73.

Appendix A

Basics of Graph Theory

A graph or undirected graph $G\ (V, E)$ consists of a set of vertices (or nodes) V and a set of edges E that interconnect a collection of pairs of vertices from V. A directed graph $G\ (V, E)$ consists of a set of vertices V and a set of directed edges E. It is represented in the same way as an undirected graph, but an arrow is placed on each of the edges in E, directing from the first node to the second node of an ordered pair of nodes. Figure A.1 shows a directed graph and an undirected graph. Note that edge (a, c) and edge (c, a) in Figure A.1(b) represent different edges.

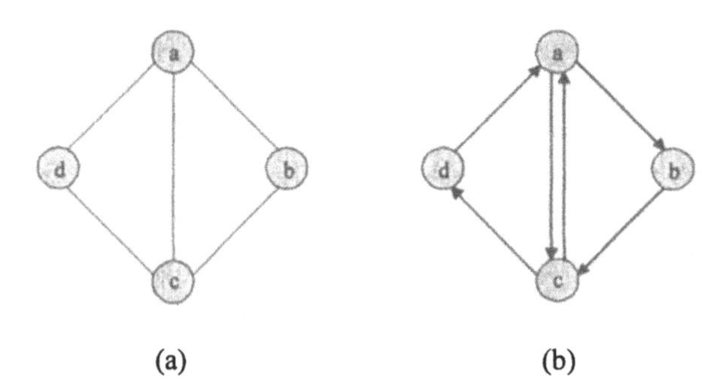

Figure A.1 Representation of a graph: (a) undirected; (b) directed.

For an undirected graph, a walk is a concatenation of edges between two nodes in the graph. A path is a walk with no repeated nodes. A cycle is a closed walk, which starts and ends at the same node with no other repeated nodes. For example, edges (*a*, *b*), (*b*, *c*), and (*c*, *d*) in Figure A.1(a) constitutes a walk between node *a* and node *d*, which is also a path between node *a* and node *d*. A cycle starting and ending at node *a* may consist of a concatenation of edges (*a*, *b*), (*b*, *c*), (*c*, *d*), and (*d*, *a*). For a directed graph, a directed walk is a concatenation of directed edges with the same direction. A directed path is a directed walk with no repeated nodes. A directed cycle is a directed walk that has more than two nodes and starts and ends at the same node with no other repeated nodes. For example, in Figure A.1(b), a directed walk from node *a* to node *c* may consist of directed edges (*a*, *b*) and (*b*, *c*). A directed cycle starting and ending at node *a* may consist of a concatenation of edges (*a*, *b*), (*b*, *c*), and (*c*, *a*) or edges (*a*, *c*), (*c*, *d*) and (*d*, *a*).

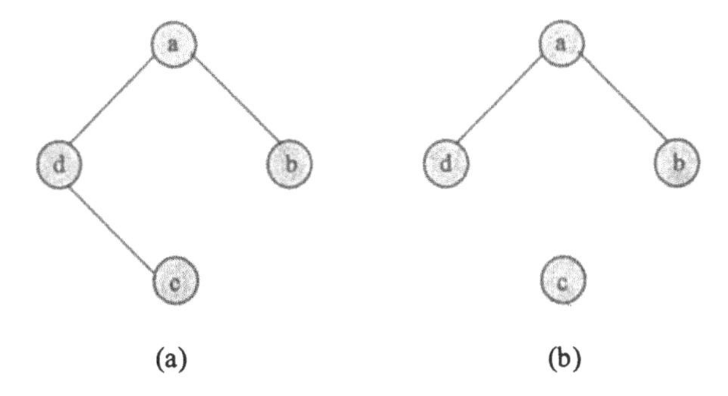

Figure A.2 Undirected graphs: (a) connected; (b) unconnected.

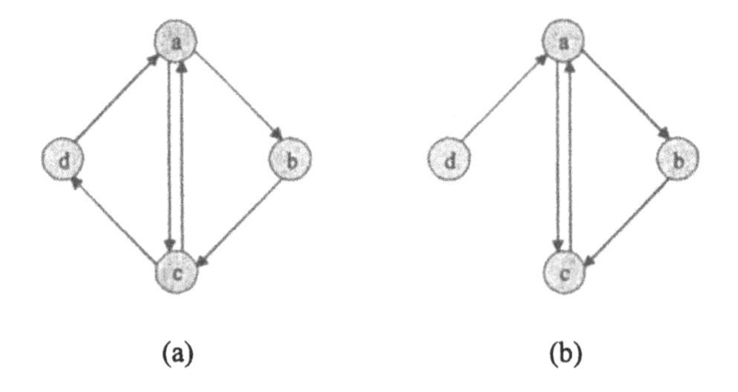

Figure A.3 Directed graphs: (a) closely connected; (b) not closely connected.

An undirected graph is connected if for each pair of nodes, there is a path between them. An undirected graph is unconnected if for any pair of nodes, there is not a path between them. For example, the undirected graph in Figure A.2(a) is connected whereas the one in Figure A.2(b) is unconnected. A directed graph is closely connected if for each pair of nodes, there is a directed path between them. A directed graph is connected but not closely connected if for any pair of nodes, there is no directed path between them. For example, the directed graph in Figure A.3(a) is closely connected whereas the one in Figure A.3(b) is connected but not closely connected because there is no directed path from node *a* to node *d*.

A graph is planar if it can be drawn on a plane such that no edge intersects with another edge except at a vertex. For example, the graph in Figure A.4(a) is planar whereas the one in Figure A.4(b) is not planar. The graph in Figure A.4(c) looks like a nonplanar graph. However, it can be redrawn as a planar graph.

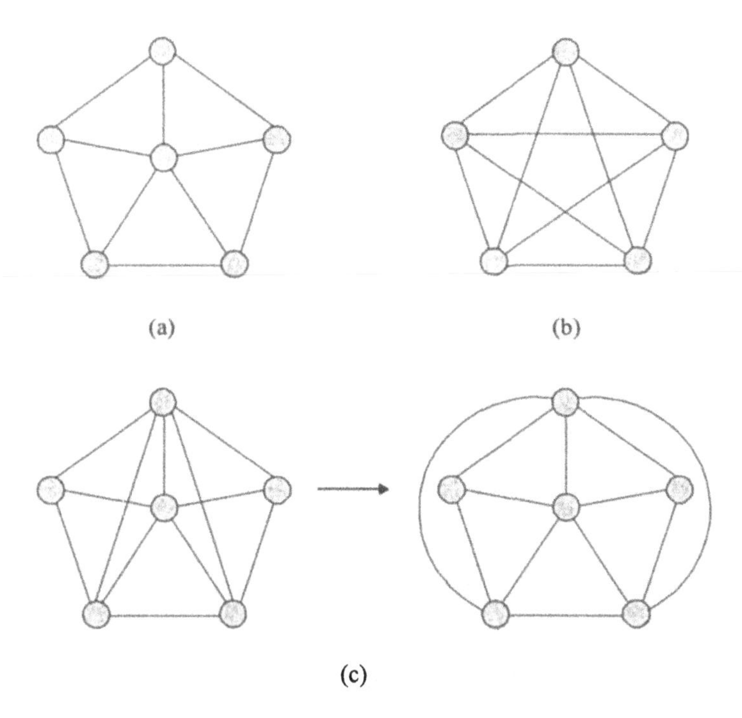

Figure A.4 Planar and nonplanar graphs: (a) planar; (b) nonplanar; (c) planar.

Two paths in a graph are vertex-disjoint if they do not share a common vertex in the graph except for their end points. Likewise, two paths are edge-disjoint if they do not share a common edge in the graph. For example, path *a-b-c-d* and path *a-g-h-d* in Figure A.5(a) are vertex-disjoint and edge-disjoint paths, whereas path *b-e-f-c* and path *g-e-f-h* in Figure A.5(b) are non-vertex-disjoint and non-edge-disjoint paths.

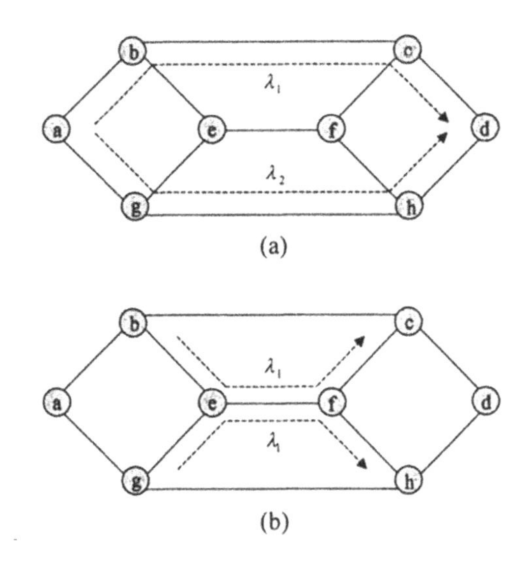

Figure A.5 Illustration of vertex-disjoint and edge-disjoint paths.

Appendix B

Dijkstra's Algorithm

B.1 Shortest Path Problem

Consider a graph G (V, E) consisting of a set of nodes V and a set of edges E. Each of the edges in E is assigned a weight, which represents the length of the edge. The length of a path is usually measured in terms of hop counts or geographic distance. Given a pair of nodes in V, the shortest path problem is to find a path with the shortest length (i.e., the minimum hop count or the shortest distance) from the source node to the destination node.

The shortest path problem has many applications. For example, if each edge in the graph represents a link in the network and the weight on each link represents the cost of using the link, the shortest path problem is equivalent to finding the least-cost path between two nodes. If the weight on each link represents the average packet delay over the link, the shortest path problem is to find the minimum delay path or fastest path between two nodes. If the weight on each link represents the congestion or resource usage on the link, the shortest path problem is equivalent to finding the least-congested path between two nodes.

There are several algorithms for computing the shortest path between a pair of source and destination nodes. In the next section, we will introduce a well-known standard algorithm for the shortest path problem, i.e., *Dijkstra*'s algorithm.

B.2 *Dijkstra*'s Algorithm

In this section, we describe *Dijkstra*'s algorithm to find a shortest path from a source node to a destination node in a graph.

Suppose that the graph has N nodes and M edges and the source and destination nodes are denoted by s and d, respectively. With *Dijkstra*'s algorithm, each edge (i, j) in the graph is assigned a nonnegative weight, denoted by d_{ij}, which represents the length of the edge. Each node is labeled with the length of the best-known shortest path to the source node, which is denoted by D_i $(i=0, 1, 2, \ldots, N-1)$. A node may be either tentatively or permanently labeled. Initially, the source node is permanently labeled. All the other nodes are tentatively labeled with a tuple (D_j, P_j) $(j=0, 1, 2, \ldots, N-1)$, where P_j is the predecessor of node j on the shortest path to the source node. If node j has no edge to the source node, the value of D_j is set to infinity. Otherwise, the value is set to d_{sj}. Then the node with the smallest D_j value among all the tentatively labeled nodes is chosen and permanently labeled. The label of each tentatively labeled node is updated. This procedure is repeated until the destination node is labeled permanently. Let $\boldsymbol{\Omega}$ denote a set of nodes already labeled permanently. The above procedure can be described in more detail as follows.

Dijkstra's Algorithm:

- Initialization: Set $\boldsymbol{\Omega} = \{s\}$

$$D_j = \begin{cases} 0 & j = s \\ d_{sj} & j \neq s \textit{ and has an edge to } s \\ \infty & j \neq s \textit{ and has no edge to } s \end{cases}$$

$$(j=0, 1, 2, \ldots, N-1)$$

- Step 1: Find $i \notin \boldsymbol{\Omega}$ $(i=0, 1, 2, \ldots, N-1)$ such that

$$D_i = \min_{j \notin \Omega} D_j$$

- Step 2: Do $\boldsymbol{\Omega} \Leftarrow \boldsymbol{\Omega} \cup \{i\}$.

  If the destination node is permanently labeled, then stop.

- Step 3: For all $j \notin \boldsymbol{\Omega}$, set

$$D_j \Leftarrow \min\left[D_j,\ D_i + d_{ij}\right]$$

  Go to Step 3.

Because there are N nodes in the graph, each step has a number of operations proportional to N and all the steps are iterated $(N-1)$ times. Accordingly, the computational complexity in the worst case is $O(N^2)$. The following example illustrates how the algorithm works with an undirected graph.

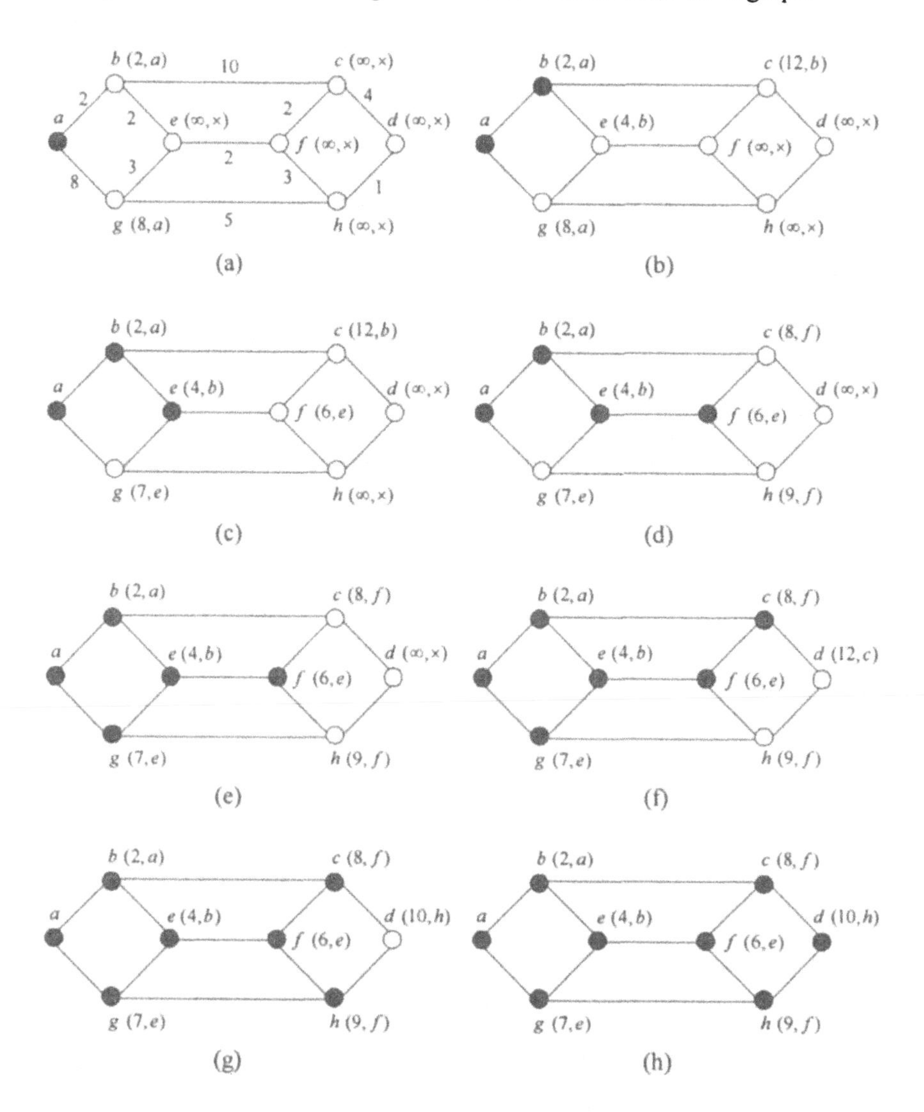

Figure B.1 Illustration of *Dijkstra*'s Algorithm.

Consider an undirected graph with eight nodes, as shown in Figure B.1(a). The weight on each link represents the link distance. We trace the procedure to find the shortest path from node *a* to node *d*.

Initially, node *a* is permanently labeled, indicated by shading. All the other nodes are tentatively labeled. Note that the symbol "×" represents that the predecessor of the corresponding node on the shortest path to node *a* has not been known yet. Because node *b* is the closest to node *a* among all the nodes tentatively labeled, it is thus chosen and is labeled permanently. Then the label of each of the other nodes is updated, as shown in Figure B.1(b). Note at this step that only the labels of node *c* and node *e* are actually updated because the other nodes have no edge to node *b*. Accordingly, the next node to be permanently labeled is node *e* because it has a shorter path to node *a* through node *b* than node *c*, as shown in Figure B.1(c). Then the labels of node *f* and node *g* are updated. Note that because node *g* has a shorter path to node *a* through node *e* and node *b*, its label should be updated. Thus the next node to be permanently labeled is node *f*, and the labels of node *c* and node *h* are accordingly updated, as shown in Figure B.1(d). Similarly, node *g* is permanently labeled, followed by node *c*, node *h*, and node *d*. Finally, the shortest path from node *a* to node *d* is found to be *a-b-e-f-h-d*, and the shortest distance form node *a* to node *d* is 10.

Appendix C

Acronyms

AAL	ATM adaptation layer
ADM	Add/drop multiplexer
ANSI	American National Standards Institute
APS	Automatic protection switching
ARBR	Alternate-link routing with backward reservation
ATM	Asynchronous transfer mode
BER	Bit error rate
BGP	Border gateway protocol
BRP	Backward reservation protocol
CR-LDP	Constraint-based routing label distribution protocol
CR-LDP-TE	CR-LDP with traffic engineering extensions
DIR	Destination-initiated reservation
DLE	Dynamic lightpath establishment
DRBR	Destination routing with backward reservation
DSL	Digital subscriber loop
EDFA	Erbium-doped fiber amplifier
FDL	Fiber delay line
FDM	Frequency division multiplexing
FEC	Forward equivalent class
FF	First-fit
FRBR	Flooding-based routing with backward reservation
FRP	Forward reservation protocol
FSC	Fiber-switch-capable
FTTB	Fibers to the building
FTTH	Fibers to the home
FTTC	Fibers to the curb

FWM	Four-wave mixing
GMPLS	Generalized multiprotocol label switching
HDLC	High-level data link control
HDTV	High-definition television
HFC	Hybrid fiber coax
HLDA	Heuristic logic topology design algorithm
HRBR	Hop-by-hop routing with backward reservation
IETF	Internet Engineering Task Force
IGP	Interior gateway protocol
IIR	Intermediate-node-initiated reservation
ILP	Integer linear programming
IP	Internet protocol
IS-IS	Immediate system to immediate system
IS-IS-TE	IS-IS with traffic engineering extensions
ITU-T	International Telecommunication Union-Telecommunication Standardization Sector
LAN	Local area network
LDP	Label distribution protocol
LED	Light-emitting diode
LL	Least-loaded
LLC	Logic link control
LMP	Link management protocol
LP	Linear programming
LSA	Link state advertisement
LSC	Lambda-switch-capable
LSP	Label switched path
LSR	Label switch router
LU	Least-used
MAN	Metropolitan area network
MAX-SUM	Maximum-sum
MILP	Mixed-integer linear programming
MLDA	Minimum-delay logical topology design algorithm
MP	Min-product
MPLS	Multiprotocol label switching
MPLmS	Multiprotocol lambda switching
MRBR	Multiple-path routing with backward reservation
MTV	Move-to-vacant
MTV-WR	Move-to-vacant wavelength-retuning
MU	Most-used
NNI	Network-network interface
NRBR	Neighborhood information-based routing with backward reservation
OADM	Optical add/drop multiplexer

OAM	Operation, administration, and maintenance
OBS	Optical burst switching
OC	Optical carrier
OOK	On-off keying
OPS	Optical packet switching
OSPF	Open shortest path first
OSPF-TE	OSPF with traffic engineering extensions
OXC	Optical cross-connect
PDU	Packet data unit
PON	Passive optical network
PPP	Point-to-point protocol
PSC	Packet-switch-capable
QoS	Quality of service
RAM	Random-access memory
RAS	Rerouting-after-stop
RCL	Relative-capacity-loss
RFC	Recommendation for comments
RSVP	Resource reservation protocol
RSVP-TE	RSVP with traffic engineering extensions
RWA	Routing and wavelength assignment
SBS	Stimulated Brillouin scattering
SDH	Synchronous digital hierarchy
SDL	Simple data link
SHR	Self-healing ring
SIR	Source-initiated reservation
SLE	Static lightpath establishment
SNAP	Subnetwork attachment point
SOA	Semiconductor optical amplifier
SONET	Synchronous optical network
SPM	Self-phase modulation
SRBR	Source routing with backward reservation
SRFR	Source routing with forward reservation
SRLG	Shared risk link group
SRS	Stimulated Raman scattering
TAG	Tell-and-go
TAW	Tell-and-wait
TDM	Time division multiplexing
TILDA	Traffic-independent logical topology design algorithm
TTL	Time to live
UNI	User-network interface
VoD	Video on demand
VPN	Virtual private network

WC	Wavelength converter
WCS	Wavelength-convertible switch
WDM	Wavelength division multiplexing
WR	Wavelength-retuning
WXC	Wavelength cross-connect
XPM	Cross-phase modulation

Index

D

E

F

G

L

M

N

O

P

Q

R

S

T

U

V

W

X

Y

Z

About the Authors

Jun Zheng

Jun Zheng received his Ph. D. in Electrical and Electronic Engineering from The University of Hong Kong, China, in 2000. Before his Ph.D. studies, he had many years of university teaching and research experience as well as several years of industry experience in developing telecommunications products. From May 2000 to August 2002, he was a Postdoctoral Fellow with the Department of Electrical and Computer Engineering of Queen's University at Kingston, Canada. Since September 2002, he has been a Research Scientist with the School of Information Technology and Engineering of the University of Ottawa, Canada.

Dr. Zheng has conducted extensive research in high-speed communication networks, covering optical networks, IP networks, and ATM networks. He is the recipient of an Outstanding Innovation Award from the Ministry of Information Industry of China for his contribution in research and development in 1995. His current research interests are in the area of optical networks, including network control architecture, virtual topology design and reconfiguration, routing and wavelength assignment, network control and management, IP over WDM, and network survivability. He has published over 30 technical papers and book chapters in this area, and has served on the technical program committees for a number of international conferences. Dr. Zheng is a member of the IEEE.

Hussein T. Mouftah

Hussein Mouftah joined the School of Information Technology and Engineering (SITE) of the University of Ottawa in September 2002 as a Canada Research Chair (Tier 1) Professor in Optical Networks. He was with the Department of Electrical and Computer Engineering at Queen's University from 1979 to 2002, where he was a Full Professor and the Department Associate Head. He has three years of industrial experience, mainly at Bell Northern Research of Ottawa, now Nortel Networks (1977-79). He has spent three sabbatical years also at Nortel Networks (1986-87, 1993-94, and 2000-01), always conducting research in the area of broadband packet switching networks, mobile wireless networks and quality of service over the optical Internet. He served as Editor-in-Chief of the IEEE Communications Magazine (1995-97) and IEEE Communications Society Director of Magazines (1998-99) and Chair of the Awards Committee (2002-2003). Dr. Mouftah is the author or coauthor of four books, 18 book chapters, more than 700 technical papers, and 8 patents in this area. He is the recipient of the 1989 Engineering Medal for Research and Development of the Association of Professional Engineers of Ontario (PEO) and the Ontario Distinguished Researcher Award of the Ontario Innovation Trust. He is the joint holder of the Best Paper Award for a paper presented at SPECTS'2002, and the Outstanding Paper Award for papers presented at the IEEE HPSR'2002 and the IEEE ISMVL'1985. Also, he is the joint holder of an Honorable Mention for the Frederick W. Ellersick Price Paper Award for Best Paper in the IEEE Communications Magazine in 1993. He is the recipient of the IEEE Canada (Region 7) Outstanding Service Award (1995). Also he is the recipient of the 2004 George S. Glinski Award for Excellence in Research of the Faculty of Engineering, University of Ottawa. Dr. Mouftah is a Fellow of the IEEE (1990) and a Fellow of the Canadian Academy of Engineering (2003).